The Suburban Chicken
The Guide to Keeping Healthy, Thriving Chickens in Your Backyard
Kristina Mercedes Urquhart
i-5
PRESS

Special thanks to my original flock of girls, Loretta, Prudence, Penny, Eleanor, Martha, Madonna, Rita, Sadie, and Yoko, for introducing me to the silly, wonderful world of keeping chickens.

The Suburban Chicken

Project Team
Editor: Dolores York
Design: Mary Ann Kahn
Index: Elizabeth Walker

i-5 PUBLISHING, LLC™
Chairman: David Fry
Chief Financial Officer: David Katzoff
Chief Digital Officer: Jennifer Black-Glover
Chief Marketing Officer: Beth Freeman Reynolds
Marketing Director: Will Holburn
General Manager, i-5 Press: Christopher Reggio
Art Director, i-5 Press: Mary Ann Kahn
Senior Editor, i-5 Press: Amy Deputato
Production Director: Laurie Panaggio
Production Manager: Jessica Jaensch

Library of Congress Cataloging-in-Publication Data
Urquhart, Kristina Mercedes, author.
The suburban chicken : the guide to keeping healthy, thriving chickens in your backyard / Kristina Mercedes Urquhart
pages cm
Includes index.
ISBN 978-1-62008-197-6
1. Chickens. 2. Suburban animals. I. Title.
SF487.U77 2015
636.5--dc23
2015023997

i-5 Publishing, LLC™
www.facebook.com/i5press
www.i5publishing.com

Printed and bound in China
18 17 16 15 2 4 6 8 10 9 7 5 3 1

Contents

Introduction

I grew up with chickens, but not in the way you're thinking. Born and raised on the island town of Key West, Florida, I rubbed elbows with all manner of feral fowl in the days of my youth. The small, scrappy birds that freely roam the island are descended from Spanish fighting cocks that were smuggled into the United States by way of Cuba. They bred with domestic laying chickens that had been freed during the postwar supermarket boom, when households starting buying their eggs instead of raising them. These wild Key West chickens still own the streets today, indiscriminately roosting in palms and brooding their babies under brush year-round, prey only to the six-toed Hemingway cats that also call the island home.

I grew up with these birds quite literally in my backyard, entirely unintentionally. They were always underfoot, roosting overhead, or, at their best, blocking traffic across town as the proverbial chicken crossing the road. My grandmother, on the other hand, having grown up in Key West as well, intentionally raised these very birds. Learning from her mother, she helped care for the flocks of chickens and pigeons that provided the family with meat and eggs. As a girl, my grandmother was entrusted with the chores of plucking the processed pigeons and egg collecting. Raising your own food was sustainable. It was healthy eating close to home. It was a way of life. And after all these years, it's a way of life that I'm aiming for as well.

Fast-forward many years and I'm 25, sitting on the couch watching television. I'm living in Brooklyn, New York, with my husband, paying all too much in rent for our tiny ground-floor apartment. I'm watching a home improvement show geared toward sustainability, taking mental notes of what I want to include in our dream house, when the homeowner leads the show's host out to the backyard. Here she has a small triangular structure on the grass with a few chickens pecking around inside. She describes her birds (naming each one, of course) and how they fertilize her soil as she moves her "tractor" around the yard.

"Ian!" I yell, though it's completely unnecessary in our 600 sq. ft. (56 sq. m) apartment, "Come look! We're going to do this when we move. We're going to get chickens!"

This idea—keeping my own chickens for eggs—excited me more than any recessed light fixture or built-in. I wasn't exactly sure what it meant, or what the larger implication was to be, but I knew I was hooked before we had even started.

When my husband and I finally mustered the gumption to quit our salaried city jobs and move out of New York, we bought our first house in his hometown in North Carolina. Erecting a coop and getting chickens was one of the first things we did in our modest backyard. We raised that first flock from day-old chicks, as we have each of our subsequent additions, and named each one after characters in Beatles' songs (Loretta, Lovely Rita, Prudence, Sexy Sadie,

Key West chickens are some of the most colorful feral chickens in America.

Michelle, Polythene Pam, among others). Feeling bold, we even named one Easter Egger Yoko, just to be contrary. She's easily the quirkiest chicken in our flock, and even now well into her prime, Yoko is still one of our best layers. Having our "ladies" in the backyard was addicting; with little effort and a short walk outside, the freshest eggs I'd ever tasted were literally within arm's reach, any time I wanted them.

It soon became apparent that keeping chickens leads to a certain way of life. What else could I procure from my efforts and my homestead? Produce? Fiber? Fresh milk? In the interest of living more fully from our land and applying a bit of elbow grease, new additions made it to our homestead each following year. Soon, we were spending our summers elbow deep in honey, tending to hives of honeybees. We filled hutches with Angora fiber rabbits, taking hours to harvest wool gently by hand. One year, we started a vermicomposting bin full of red wiggler worms to turn kitchen waste into rich fertilizer. We blamed this new lifestyle almost entirely on the chickens; they had been the catalyst for changing the way we saw food and other basic products that we were taking for granted. Without too much exaggeration, you might say they encouraged us to move away from New York. Soon, we had dubbed chickens "gateway livestock" as they had opened the door to this ever-growing menagerie of furry, feathered, fuzzy, and buzzing charges that changed the way we ate.

Most of the homestead additions had slowed by the time our first child was born, though. Her first bites of solid food came from squash grown in our soil; her first (and only) eggs were from our flock of chickens (thank you, ladies). With our daughter's presence and increasing involvement around the homestead, it became very clear that we were doing everything we were doing for her and for the next generation of stewards of the land. My hope is that our daughter, like my grandmother, will grow up with respect for the animals that provide her with food and with a hands-on experience that garners the deepest gratitude for those beings.

There is a revolution happening in America, folks. It's taking place in backyards all across our country's cities and suburban neighborhoods. Americans are reclaiming the rights to which foods make it onto their plates. They are recognizing their inherent freedom to eat healthy foods, to know where it comes from, and most importantly, to teach the next generation that they can eat well and feed themselves, with their own two hands, by caring for the land. And in my opinion—and experience—that begins with chickens.

A feral hen and rooster pair roam the streets of Key West, Florida.

Chapter 1

Why Keep Backyard Chickens?

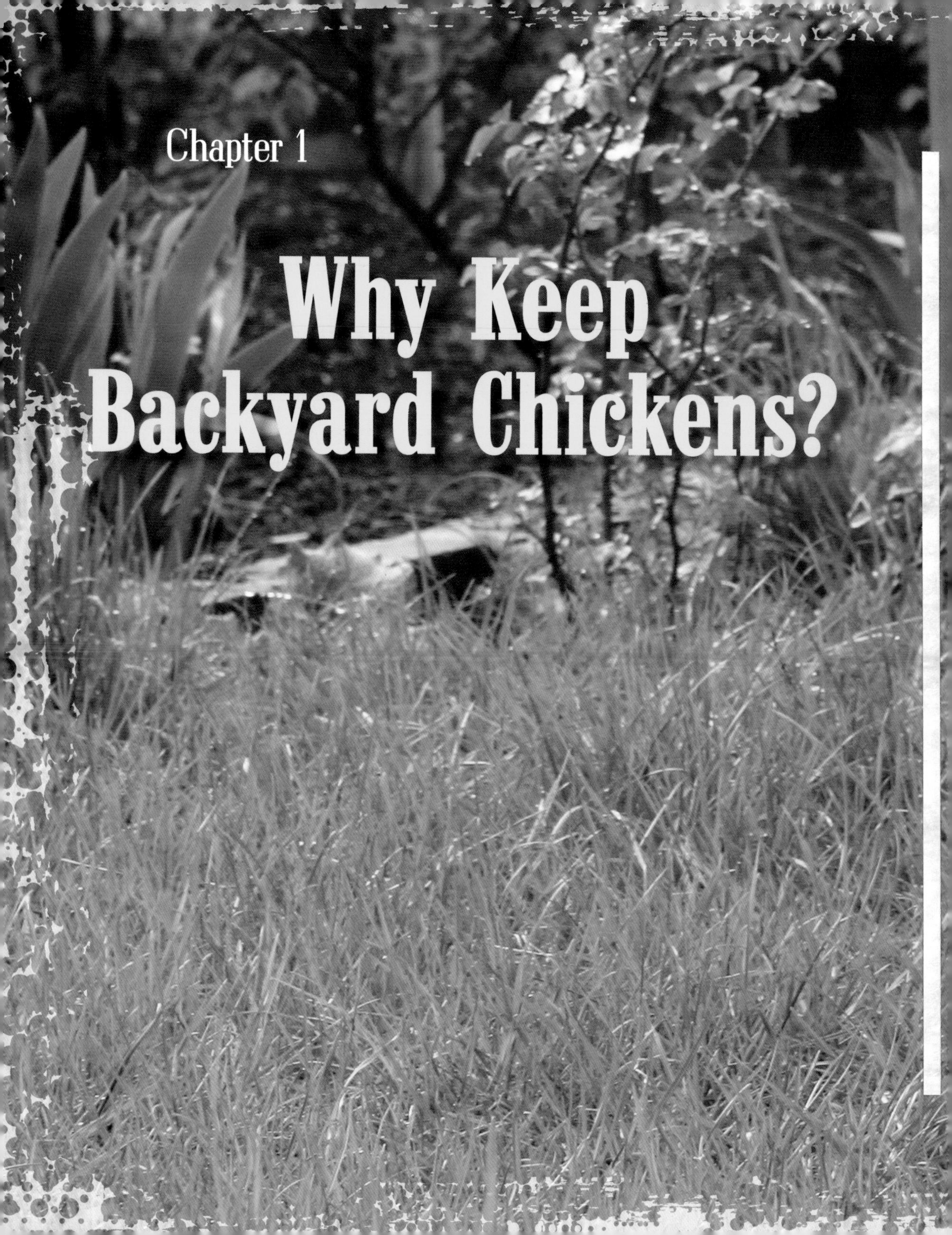

Since you're holding this book in your hands, you're likely already aware of the chicken revolution taking place in backyards across the country. Homeowners, renters, families, couples, college graduates, nine-to-fivers, aspiring farmers, artisans, and countless others are cramming nest boxes onto balconies and turning idle potting sheds into chicken coops. Each year, more and more cities are overturning ordinances and allowing chickens to come home to roost within their limits. There's no doubt about it: There's a bona fide chicken craze underway.

Who Is the Backyard Chicken?

The *Why* of keeping chickens is probably better answered by asking the *Who* question. Who is the chicken? This humble ground fowl, the most iconic of American farm symbols, wears many hats, whether she lives on 50 sprawling acres (20 ha) in the country or in a 50 sq. ft. (4.6 sq. m) backyard in the suburbs. She was actually once rather common in American homes. Most families kept a few laying hens until the convenience of supermarket eggs simplified the householder's duties, and small flocks were replaced by Big Agriculture. The chicken is, in fact, the cornerstone of any good permaculture system, and she is becoming the very foundation of awakening our flawed food system. By learning a bit about who she is, you may come to understand why so many are becoming smitten with her endearing ways.

A Prehistoric Being

Gallus gallus domesticus is first and foremost a bird. More specifically, she is part of a class of ground fowl that includes other domesticated poultry such as turkeys. Likely descended from the Asiatic red jungle fowl, recent evidence suggests that she is the infamous tyrannosaur's most closely living relative alive today.

This would be reason enough for me to keep chickens (tiny T Rex roaming in my backyard? Sign me up!), but it's probably not convincing enough for most. Let's see what else she can do.

Get Hooked on Breakfast

Homegrown eggs are not just any eggs—they are the healthiest, freshest, tastiest ones you will ever have, hands down. The vast majority of chicken keepers start a flock for these incredible eggs, and it's no wonder. The eggs laid from backyard flocks that forage for part of their diet are higher in omega-3 fatty acids, vitamins A and E, and beta-carotene than store-bought eggs. They're also higher in folate (the naturally derived version of folic acid) and vitamin B_{12}. Pastured eggs are lower in saturated fats than their supermarket counterpart and have nearly half the cholesterol. They are humanely sourced, since you are raising the chickens yourself, and they are as local as it can possibly get.

All female chickens lay eggs, but through generations of selective breeding, breeders have homed in on birds that lay exceptionally well, creating the great laying breeds we have today. Some breeds are veritable egg-laying machines, cranking out an egg or so a day for the first year or two, with the frequency dwindling as the hen gets older. Among this class of workhorses are docile birds that have great personalities and make wonderful family pets. Despite some hens being better layers than others, all hens lay eggs with varying frequency—and contrary to some misconceptions, hens do not need a rooster to lay eggs (but you will need a rooster if you want the eggs to be fertilized for hatching). Under the proper conditions and with dark, dry, soft, and cozy nest boxes, your chickens will continue to lay eggs well into their twilight years.

Keeping chickens will give you fresh eggs that are more nutritious and have less cholesterol and saturated fat than store-bought.

This is where you'll get hooked. Warm eggs, freshly laid and sourced straight from your backyard, back deck, or patio cannot be beat by any measure of the imagination. Watching your hens lay their first eggs is nothing short of witnessing the miracle of life. But tasting them? Well, that can be downright heavenly.

Better Eggs, Fewer Bugs (the Natural Way)

Some large-scale egg producers like to tout that their hens are "fed a vegetarian diet only" on their cartons' packaging, but in reality, chickens are anything but vegetarians. They are, in fact, ruthless hunters, with worms, grubs, and other insects high on their list of menu choices. Chickens will even chase and kill small rodents and snakes if given the opportunity. They really are like tiny dinosaurs. More common, though, is their ability to forage on grass, in gardens, and in pastures for their choice of seeds, greens, and bugs. This variety in diet is what contributes to saffron-colored yolks and those ultra-nutritious eggs mentioned earlier. Simply put, variety in your birds' diet converts to nutritional variety in your eggs. The healthier your birds are, the healthier your eggs will be.

Nutritious eggs aren't the only reason to let your birds range freely, though. When left to their own devices, chickens will scratch and peck at the earth until they have foraged for every last kind of bug—especially garden pests,

The Great Recycler

We know that chickens will eat just about anything, and as such keeping chickens means you have a reliable team of enthusiastic garbage disposals at the ready. Your flock will contentedly gobble up everything from cooking scraps and kitchen leftovers to grass clippings and spent garden plants. While turning over your compost piles, they'll eat any vegetation, fruits, or vegetables that aren't too far past their prime. Chickens aren't very picky, but, because no creature is infallible, there are, of course, some toxic foods and poisonous plants that can be harmful to chickens. We'll cover those, and more on food and nutrition, in chapter 7.

such as slugs, Japanese beetles, cabbage worms, and those that carry diseases, like the common deer tick. Even mosquitoes make the list. Just your chickens' presence around your home will keep pests such as mice, rats, and snakes at bay. And they do all of this without the use of harsh chemicals, pesticides, or toxins. Chickens are, without debate, the most efficient, natural insect and pest control there is.

The Fertilizer and Garden Helper

The chicken is an excellent producer of chemical-free, organic fertilizer. Long past her reproductive prime, a chicken will continue to produce manure in copious amounts (for better or worse), even when she has slowed down laying eggs. High in phosphorus and potassium, but especially in nitrogen—the critical elements for healthy, fertile garden soil—chicken waste is one of the most perfect manures of the animal kingdom. It is easy to harvest from the coop for use in composting, and your flock will naturally spread it on the land as they range freely or forage in movable coops.

Chickens will also happily aerate your compost pile through their pecking and scratching. Just make sure they have access to it and that there are no dangerous chemicals or garden waste that have been treated with pesticides, fungicides, or synthetic fertilizers in it. Not only are these dangerous for your birds, but these toxins could also pass through their bodies, into their eggs, and into your breakfast.

Chickens make quick work of clearing weeds and grasses if you need bare soil for a garden space or other project. Just fence your birds in that spot for a week or more, and they'll happily demolish any greens, leaving behind aerated, fertilized, and gently tilled topsoil ready for planting.

Your Insurance Policy

Keeping your own flock of laying hens means you have your very own sustainable and reliable source of food, no matter what may be happening in the larger food or transportation industries. Our food travels far and wide to get to us; any disruption to the trucking industry can mean no food on marketplace shelves. Droughts, cold winters, or heat

waves can mean food shortages or skyrocketing prices, too. Despite product recalls or salmonella outbreaks, for example, you know you can rely on a safe source of food for your family, direct from your own backyard. With a substantial flock of laying hens, you may even have a source of income from selling eggs to friends and neighbors.

The Low-Maintenance Pal

We all have one. Someone we're so comfortable being around that we can really be ourselves—no pretenses, no judgments. The chicken is that kind of friend.

Sure, she may be flighty and unpredictable at times, but mostly she's funny, chatty, quirky, and endlessly entertaining. One of my favorite ways to spend a lazy summer evening is in the backyard, cool drink in hand, watching my flock peck around the yard, perch on my lawn chair arms, investigate my flip-flopped toes, or coo when they find something tasty in the grass. Each bird has a distinct character, and it shows. Each has a distinct personality, status in the group's pecking order, food preferences, roosting space, favorite dust-bathing spot or nesting box, and so much more.

If you don't care for long walks, litter boxes, or dog parks, you've found the perfect pet. An entire flock of hens is easier to care for than a single housecat, and they're far less temperamental. With certain methods and styles of housing, coop maintenance can be a breeze. Daily watering and feeding are quick and easy, and when left to forage most of the day, a flock can easily find much of its food on its own. Once you make the initial investment, a flock's upkeep is inexpensive and doesn't take much time at all. By taking preventive measures and paying attention to cleanliness, chickens will rarely fall ill and can remain spunky, healthy, animal friends for up to ten years or more.

This Black Australorp forages in a compost pile. When properly composted, chicken droppings make a great addition to your garden.

The Educator

The chicken reminds us that our food is a living thing. The animal that produced that egg, meat, or dairy product was a living creature. The plant that grew the produce on your plate lived in soil and swelled under the sun (unless it was grown hydroponically or in a greenhouse). These food products came from living things that were grown, raised, harvested, and cared for by other living things—people, farmers.

This may seem obvious, but too many of us have lost touch with the origins of our food. In our busy day-to-day lives, it is easy to disconnect from these basics. The chicken can remind those of us who have forgotten and teach those of us who are just learning.

The chicken is the mascot for the back-to-the-land movement for good reason. She teaches us that food is precious and should not be mass-produced. Mass-produced food and food products lack nutrients, integrity, and heart and soul. Recent recalls of factory-farmed eggs have shed light on an industry stealthily growing ominously right under our collective noses. The evidence is contained in the white foam cartons stacked beneath fluorescent lights on the grocery store shelves. The hens that laid those eggs were crammed six or seven to a cage the size of a standard sheet of paper, their beaks trimmed to keep them from cannibalizing their compatriots in the cramped quarters where they pass their lives. Once "spent," they are trucked off to become broth or soup for another industry that considers this living being a "product."

Sayings from folklore have been passed from one generation to the next for thousands of years. Many have a certain rhythm and are most often rooted in explanations of things happening in the natural world. Despite centuries of change, many of these, like the familiar chicken sayings below, remain in the modern lexicon.

As scarce as hen's teeth: to be difficult or impossible to find; a rarity.
Mother hen: a matronly mother figure.
Nest egg: to save a bit of money over time.
Chicken scratch: a small amount of money.
Henpecked: to be picked on or nagged.
Don't put all your eggs in one basket: a warning against investing all prospects or resources into one venture at the risk of losing everything.
To be chicken: to be afraid or cowardly.
Walking on eggshells: the need to be careful or tread lightly around the feelings of another.
Something to crow about: to have exciting news to share.
Have egg on your face: to be caught in a lie; to be guilty of a shameful act.
Fly the coop: to have left home; to be gone.
Like a chicken with its head cut off: to have no direction; to flail, literally or figuratively.
Made from scratch: handmade; to create something using raw materials.
Empty nest syndrome: a feeling of sadness, loneliness, or melancholy after grown children leave home.
To brood over: to worry or fret over a problem or situation.
No spring chicken: an old, tired person.
Shake a tail feather: to get a move on; hurry up.
To ruffle a few feathers: to annoy some people when making changes or improvements.
Birds of a feather flock together: those with similar personalities or attributes befriend and spend time with one another.
Not all it's cracked up to be: when a person, thing, or event does not live up to expectations, reputation, or popular hype; a disappointment.

As if that weren't reason enough to bring your dollar closer to home, those eggs are not what they could be. Without variation in their diet, eggs from factory-farmed hens plummet in heart-healthy omega fatty acids and soar in "bad" cholesterol. And because these eggs have to travel far and wide to get to thousands of supermarkets across the country, they are often several weeks to a month old by the time they reach your shopping cart.

Raising chickens can be fun and very rewarding, too. They are easy to look after, don't take up much room, and provide a steady supply of fresh, nutritious eggs.

Keeping backyard chickens offers a valuable opportunity for education—not just your own but also for your children, partner, friends, family, neighbors, and greater community—about the realities of factory farming and the industry we've been inadvertently supporting with our dollars. Keeping backyard chickens ensures that your birds are cared for in a humane and respectful way and that they live dignified lives, free from unnecessary suffering. In turn, you'll be rewarded with the healthiest eggs available. Erect a beautiful coop that you're proud of and invite the neighborhood over for a coop-warming party. Or, take your kid's pet chicken to his class's show-and-tell day and empower him to educate his peers. Surprisingly, there are many who just don't know that a chicken lays an egg and that an egg is not just a manufactured food product that turns up on grocery store shelves. Or, that it's not a dairy product. See? As a culture, we have collectively forgotten some of these very basic tenets of living.

Through your chicken keeping, you can educate a community about where their food comes from. You and your family can show friends and neighbors that chickens can be respectfully cared for, live happy lives, and still provide you with more eggs than you can eat.

Who Is the Chicken Keeper?

So, that's the chicken. But who is the chicken keeper?

The answer is simple: anyone.

The chicken keeper is someone who values sustainable living, likes good food, wants to take control of his or her food sources and personal health, and enjoys the company and entertainment of chickens. The chicken keeper is someone who has a bit of outdoor space to spare for a flock (they're social animals, so they need companions—a bare minimum of three chickens is best). The chicken keeper is someone who has time for quick daily feeding and watering, basic monthly maintenance, and a yearly coop clean. The chicken keeper doesn't mind occasionally getting his or her hands dirty and doesn't mind sharing garden space with a few quirky birds.

Chapter 2

Keeping Chickens in the 'Burbs

In my grandmother's youth, a small family flock in the suburban yard was commonplace. Chickens clucked contentedly in the background as American families went about their daily lives. It wasn't a hobby so much as a way of life. In tandem with a garden, putting food on the table directly from the land was how it was done. After World War II, a massive shift in the way Americans eat began to take place. With the advent of highway systems and the growth of cities, food was more easily transported, and grocers could stock their shelves with exotic foods grown some distance away. Eventually, food was mass-produced on large farms, frozen, prepackaged, and trucked to local grocery stores rather inexpensively. Acquiring food in this way soon became incredibly convenient for the typical American family. Trade artisans, such as bakers and butchers, began to disappear, as it became easier to stock a kitchen or pantry with everything from under one retailer's roof. And so, the American supermarket was born.

Almost simultaneously, many suburban and rural communities created regulations against the keeping of domestic chickens. Some cities outlawed the keeping of chickens within their limits altogether (although others, such as New York City, never changed regulations and still allow backyard flocks to this day). With the resurgence of self-sufficient philosophies and values and the reclassification of chickens as pets rather than livestock, many cities are now overturning their ordinances and allowing backyard chickens once again.

Building an attractive chicken coop may not matter to the chickens, but it will go a long way in promoting good relations with your neighbors.

Are Chickens Right for You?

Almost anyone can raise chickens, but is keeping chickens right for you? Whether your city or town has recently overturned ordinances or never shed its rural roots to begin with, this chapter covers all of the concerns, costs, and considerations you need to think about before embarking on backyard chicken ownership.

Your City's Rules and Regulations

Like anything worth doing, there may be a few hoops to jump through before you can start your flock. First, it's important to determine whether your city or town is one that allows backyard chickens.

Start by contacting your city's health and zoning boards to see if chickens are legal within the city's limits. If you're lucky, your town may have municipal codes posted online for the public to read. (Be forewarned: Municipal codes are dry reading and can be a tad bit overwhelming.) Look under headings labeled "Animals" and "Zoning"—these two sections will likely provide some answers. If your online browser offers a search function, try searching for words like "fowl," "poultry," "livestock," and, of course, "chickens." Using these keywords to find the appropriate paragraphs could significantly speed up your search time.

When in doubt, county clerks or code compliance officers are great resources to help you sift through the heady terminology. However, if you speak with someone who assumes chickens aren't allowed in your city, ask to see the codes that state the rules to be sure. For some municipalities, there may be no mention of poultry under the "Animals" section, but that doesn't mean they aren't allowed. Sometimes, the rules for how many of each animal species is legal per zone may be under "Zoning," so try searching there before you give up. If your city *does* allow chickens, bear in mind there may be a limit to the number of birds you may keep, and there may be additional fees for any birds or other pets over that limit. Some cities may also require a permit and charge a small yearly fee.

Finally, if you live in a neighborhood with a homeowner's association, you'll also want to check with the association's board to learn whether they allow chickens in your neighborhood. If they are against chickens, you may feel passionate enough to take steps toward reversing those restrictions. Educating your fellow homeowners and board members is the first step: Speak to your neighbors, start a petition, and present to the board all of the many reasons why chickens are a great addition to any backyard.

Know Your Chicken L.O.R.E.

To find out about your city's chicken laws and ordinances, go to the website backyardchickens.com. There, you can search by state to learn all about chicken L.O.R.E. (Laws and Ordinances and your Rights and Entitlements): www.backyardchickens.com/atype/3/Laws

Being Neighborly: Proper Chicken-Keeping Etiquette

So your city's codes check out, and you're legally allowed to keep chickens in your town. Hooray and congratulations!

Before you start ordering chicks and building a coop, though, you may want to run your intentions by your neighbors. Those not familiar with chickens may be wary of even the mention of the barnyard animals. After all, chickens make noise (just a little) and their housing can smell (if improperly cared for). Your

The Right to Bear Chickens: Eight Tips for Changing City Ordinances

If you find that keeping chickens is illegal in your hometown, all hope's not lost. It is possible for citizens to petition, educate, and lobby to have an ordinance overturned. Really, anyone can do it. If you're passionate about poultry and are ready to see legal backyard chickens in your city, here are eight tips for getting started.

1. **Knowledge Is Power.** Keep reading this book—and any other chicken book you can get your hands on. To convince fellow citizens, skeptical neighbors, and government officials that chickens belong in backyards, you'll need to know enough about them to speak confidently about their needs, behavior, and proper care. Get to know chickens inside and out by visiting regional farms and speaking to current chicken owners in neighboring towns or cities.

2. **Build a Team.** Next, find like-minded citizens who are willing to help see this project through to the end. Assemble a small task force that includes two to four friends, neighbors, or backyard chicken enthusiasts to help you. These should be individuals who are familiar with chickens and feel comfortable speaking to groups of people, have the time to commit to the project, and with whom you feel comfortable collaborating. Other community allies may include neighborhood associations, master gardeners, local and slow-food advocates, schools, senior centers, animal rights groups, local chefs and restaurants, farmers, and food banks.

3. **Build an Image.** Give your movement a name. Then, with your teammates, start a mailing list of supporters and get the word spreading. Create a website, blog, or social media group (through sites such as Facebook and Yahoo), and invite citizens and local businesses to join, offering updates throughout the process and asking for help and support as needed. Getting endorsements from the community is important. It's easy to ignore just one or two people who want to keep chickens—it's hard for city government to ignore whole neighborhoods, vocal families, and local businesses.

   Also, organize meetings for your supporters. Offer free screenings of films such as *Mad City Chickens* or *Chicken Revolution* for inspiration and encouragement. Find an existing chicken keeper who has a particularly docile hen that might be enlisted as a goodwill ambassador. You'll be amazed how eager people will be to sign petitions, offer support, and join the project when they come face to face with a living, breathing, fluffy hen.

4. **Get into Government.** Educate yourself on how city government works and get to know your city officials, planning staff, or advisory board. If you can, identify individual members who may have a soft spot for the project and will take you under their wing. Having a pro-chicken ally within the city government can be extraordinarily helpful and speed up the process significantly.

5. **Learn from Others.** Learn the laws from other chicken-friendly towns and offer these as examples to your city council. What do neighboring cities' regulations on keeping chickens look like? How many birds are allowed per residence? How far must the coop be placed from existing structures? Do your homework and come armed with information; you'll inevitably need to navigate (legitimate) concerns about noise, smell, and curb appeal from citizens and council members. Be prepared to answer questions.

   Also, find out how successful backyard chickens have been in those cities. If the movement is flourishing and complaints are at a minimum, you may ask for a letter of support from those cities to take to your council members. If you're particularly savvy about governmental affairs, you may even draft a proposed ordinance for keeping chickens in your town (modeled on those cities' codes) and offer it at a meeting.

6. **Get the Media Involved.** Contact your local newspaper and other city publications that may be interested in covering your story by targeting reporters who have an interest in sustainability and environmental issues. Also, accept offers to speak on the radio and similar public platforms. Be outspoken and spread the word. The goal is to garner widespread support and to put pressure on your city council to address the issue (they'll be more likely to pass the ordinance if they see a majority supports it). Get the conversation going. Any press is good press.
7. **Come Prepared.** With the information you've gathered about chickens and other cities' chicken-keeping laws, put together information packets to give to the city council member(s) who are willing to help you. If no one is willing, attend town hall meetings any time chickens are on the agenda. Gather your supporters and ask them to turn out to show support. Pick a few eloquent individuals to prepare speeches on certain concerns (odor, noise, disease, and so on) and ask them to speak when the council is open to public comments. Make your voices heard but be polite and stick to the facts. Show council members that you are serious and determined to see the project through. And don't forget to invite the media.
8. **Be Patient.** Some cities are able to overturn ordinances in 6 to 9 months, but many take 12 to 18 months to see results. Once you get the ball rolling, don't give up. Remember, "The squeaky wheel gets the grease" so be persistent but also compassionate and courteous. Do your research, be reasonable, and respectful, and you'll soon have legal chickens in backyards all over town.

job is to consider your neighbors' concerns and alleviate any unfounded fears. Assure them you'll keep a clean coop (and follow through on that promise), because a well-maintained coop doesn't smell. Also educate them on basic chicken behavior: Many people don't realize that hens do not need a rooster to lay eggs, so you won't need one around. (Most cities prohibit roosters, anyway, and roosters are the noisy ones.) Hens cluck and chirp contentedly throughout the day, but rarely do they make noise above a conversational speaking voice. Occasionally, an individual bird may sing her post–egg laying "hen song," but it only lasts for a few minutes and is still quieter than a barking dog.

Finally, take aesthetics into account. A nicely constructed coop with attractive paint choices and window boxes or some minor landscaping can go a long way toward winning over wary neighbors. Keep it clean and tidy. Remember, your coop represents the modern chicken movement, and you are a representative of chicken keepers everywhere. Show your neighbors and prove to your town that chickens do have a place in every yard.

And when all else fails, shamelessly bribe your neighbors with the enticement of homegrown, pastured eggs with yolks so dark, they're nearly orange. That might just be enough to win them over.

A Lifestyle with Chickens

Any new addition to a family requires some adjustment, and adding a flock of chickens is no different. The good news is that with a little preparation, chickens can fold neatly into nearly any lifestyle, any schedule, and any backyard or neighborhood.

Like any pet, caring for chickens requires a bit of daily attention. Whether you have a flock of 3 or 30, daily tasks include egg collecting,

Chickens are among the easiest of animals to care for, but that doesn't mean the birds are care free.

visual checks on all of your ladies for illness or injury, and a quick scan of the coop's perimeter. That's it. If you provide your flock with large water fonts and feeders, feeding and watering chores can happen every other day or so, and big cleanings can be reserved for one weekend in the spring.

One of the biggest adjustments to your lifestyle will take place in the evening, when the flock turns in for the night. Chickens instinctively seek shelter at dusk, and they'll find it in their coop. To keep them safe from predators, they'll need someone to lock them up in the coop each evening and open the coop door again each morning at daybreak. This chore isn't time-consuming in itself; rather, it's all about timing. Many predators strike right at dusk or shortly after, so it's important to close up the coop just after the flock retires for the evening. The time for sunup and sundown will gradually shift with the seasons, though, changing the time of day you'll need to close and open the door. For these reasons, it helps to create a system in place that works for you. Some chicken keepers install automatic coop door openers that close on a timer. These can be costly but really convenient. Some chicken keepers let their birds free-range only when they're at home to watch them and are available to shut the door behind them each evening. Others build predator-proof outdoor runs that don't require the daily opening and closing of the coop door at all (very helpful for those who work late or have unpredictable schedules). For those of you with families, this is a terrific responsibility to give to children and teens.

Another consideration when starting a flock of chickens is establishing their care while you are away. Unlike cats or dogs, a flock of chickens can't be boarded with a veterinarian or kennel. Whether you travel for work or vacation, you'll need to employ someone to come to your coop to gather eggs, feed and water the flock, and open and close the coop door while you're away. Fortunately, as the chicken-keeping movement gains momentum in cities across the country, more and more knowledgeable chicken sitters are offering their services. For a small fee, they'll keep an eye on your flock and know what to look out for should a problem occur.

Finally, if any of these responsibilities seem burdensome or an ill fit for your lifestyle, consider sharing the weight with others. Whole neighborhoods have been known to get in on the action of keeping chickens, with great results. Starting a flock with neighbors helps to distribute the costs and responsibilities of keeping chickens, and there will always be extra sets of eyes, ears, and hands to pick up chores, troubleshoot issues, and swap egg recipes.

Expenses

Many assume that if they keep a few chickens, they'll get "free" eggs. But if you're doing it right, keeping chickens is not free. While day-old chicks are rather cheap (cheep!) compared to other pets at $3 to $5 per chick, there will be start-up costs in the way of housing, equipment, bedding, feed, and supplements. The latter three items will be continuous monthly or yearly costs as well, so budget accordingly. To some degree, spending money is necessary if you want to keep healthy chickens.

By far, the coop is often the largest expense. You can spend thousands of dollars on a deluxe chicken coop with all the bells and whistles, or you can get creative and retrofit an old shed or doghouse with found or recycled materials—or something in between, of course. It's really up to you, your budget, and your skill set. As long as the coop keeps your birds safe from predators and the elements, and it's strong, sturdy, and gives them a place to lay eggs and roost at night, they won't mind what the coop is made of or how much it costs.

You can spend as much or as little on a coop as you'd like.

Your chickens will need a coop that properly protects against predators as well as inclement weather.

Like any pet, chickens require a continuous supply of feed that will accrue a regular (usually monthly) cost. The thought of chickens foraging in the backyard (for free) is certainly idyllic, but it's not realistic. Laying hens have specific nutritional needs, and if they aren't met, their health will surely suffer. Feed costs vary widely depending on the brand of feed you choose and the number of birds you keep. (There's much more on feed and supplements in chapter 7.)

The hard lesson that many excited, new chicken keepers learn in the first few years is that you don't really save money on eggs by keeping chickens. If you've been buying pastured, local eggs from the tailgate market or nearby farms, you likely have an idea of what eggs from healthy, humanely raised chickens actually cost. In that case, you may see a small savings. But if you're transitioning from 99¢ per dozen for factory-farmed eggs bought at the grocery store, it may be several years before you start to see a difference in your food bill. Even so, many keepers, myself included, feel that when the true costs are weighed, the effort is well worth the price.

Space Requirements

Chickens need room to roam. How much, exactly, depends on how large your flock is. The jury is still out on precise numbers per bird, but a good rule of thumb that many chicken keepers follow is to provide 4 sq. ft. (0.4 sq. m) of coop space per bird if they free-range daily and 10 sq. ft. (0.9 sq. m) of space per bird if they are confined to a coop, pen, or outdoor run full time. You can't really give a chicken *too* much space; in this area, more is definitely better.

The only hard-and-fast requirement is that the space be located outdoors. To be a chicken keeper, a backyard space is a must. A small flock could conceivably live indoors and have their basic needs met, but I wouldn't recommend this route. Raising chickens indoors would deprive them of their very nature to scratch, dig, peck, and dust-bathe. And giving laying hens a better life is one of the reasons we get into chicken keeping in the first place.

Time Requirements

We've touched a little on the basic time requirements needed to tend to a small flock of chickens. There will be a few (very easy) daily chores but also some (moderate) weekly maintenance, and once a year or so, there will be a (rather messy) coop cleaning. For now, be prepared to devote some time to your coop's maintenance and chicken care.

Chores aside, when you share a life with chickens, there will be some lost time that goes unaccounted for: watching baby chicks in a brooder, spending time in your garden as your flock follows you around, and watching young pullets lay eggs. These are the "chores" that make all of the poop scooping worthwhile.

Kids are fascinated by animals, so it's only natural they will want to help tend the chickens.

Peeps, Pipsqueaks, and Poop: Kids and Hygiene

Chickens and kids can—and often do—get along famously. Many of the small chores needed to properly care for chickens are perfect for little hands. Children are often endlessly entertained by chicken antics, and raising chickens for eggs offers valuable insight into where our food comes from.

To make the most of this relationship, though, children will need adult supervision in several areas. The most important factor is hygiene. Chickens aren't the neatest and tidiest pets there are, and anyone tending to them will become quite accustomed to navigating around their waste. Proper hand washing should follow any contact with chickens. Wearing appropriate chicken attire (such as muck boots) for anyone working or spending time in the coop is also a must. By creating a hygiene routine for children directly involved in chicken care, everyone can rest assured knowing the poop stays outside where it belongs.

Dogs and Chickens: Know Your Pooch

Can dogs really be trusted around chickens? You've likely heard a few disaster stories: traumatic events that ended in a bloodbath with dead chickens and Fido to blame. Others swear by their pooch's loyalty, and you've no doubt seen the social media photos of placid, sleeping canines with fluffy chicks happily perched all over them. But when it comes to chickens and dogs coexisting in reality, there are no hard and fast rules. The real answer is that every dog is different.

Chicken breeds, while they vary widely in color, shape, and sometimes personality, are relatively uniform when it comes to temperament. They are creatures of prey, and they don't want to be eaten, so they'll do everything they can to avoid that fate. This makes them pretty predictable in terms of how they will react around most "predator" animals (dogs and cats are included in this category). On the other hand, the personality and temperament (not to mention size) of many dog breeds, in combination with a dog's individual personality, can vary so dramatically that only you will be able to determine if your pup can be trusted around chickens. Here are some of the different factors that contribute to your dog's trustworthiness around prey animals.

Working with Breed

Before you can move toward any conclusions about your dog's behavior (or potential behavior), know his breed. If he is a mix from a shelter, then make your best guess. If you're completely stumped, consult a canine trainer or dog expert for his or her advice. Knowing your dog's dominant breed and the general traits of that breed is important because it will tell you a lot about what to expect of your dog around smaller animals. Read about the breed and what purposes or tasks it was developed for. Was your dog's breed cultivated for herding? Hunting? Companionship? These offer very important clues to your dog's underlying temperament and will help inform whether he or she will get along with chickens.

The reason this is important is that some breeds have a relatively high prey drive. Prey drive is the tendency or inclination to attack or pursue, and sometimes kill, smaller animals that are perceived as prey. Prey drive is not the same thing as aggression. Some breeds have a lower or higher prey drive depending on what they were bred to do. It is possible to train a dog around its naturally inclined prey drive, but this takes time, commitment, and, above all, consistency from the dog's owner.

Working with Personality

Even a dog of a passive, unassuming breed can still pose a threat to your chickens if its personality is aggressive or overly playful. My 90 lb. (40 kg) retriever mix would never intentionally harm one of our birds, but she is playful, very large, and doesn't know her own strength. While playing, she could easily pounce on a chicken and inflict a fatal wound, even without harmful intentions. On the other end of the spectrum, one of our Chihuahuas (we have two) seems to believe that everything, and everyone, exists to play with her. She likes to chase the chickens around in fun, but they find it rather annoying. They sometimes chase her back but simply aren't fast enough to catch her. The dog doesn't pose a threat to the chickens (if you don't count sheer annoyance), but this behavior under other circumstances could stress out certain birds, separate chicks from hens, or make them feel generally unsafe. In this example, a breed developed for companionship (the Chihuahua) is overshadowed by the dog's personality (high energy, very playful, and occasionally focused on the chickens). Sometimes, individual dogs simply defy their breed; their personalities just don't match what their breed says they should be like. This is where it's up to you to know your dog and watch his or her signs around perceived prey. Again, it all depends.

Birds of a Feather: Making Introductions

So you've determined your dog's breed, know his personality, and think he's ready to coexist with chickens. Where do you start?

First, timing matters. If you've recently added a puppy to your household and already own chickens, introduce them right away (as in, while your dog is still a puppy). If you already have a dog and are considering getting chickens, you may want to test your dog's prey drive and general reactions by introducing him to other birds before getting chickens, with the supervision of a professional, of course. The following method is roughly the same for both scenarios.

The safest option is to start with the least amount of exposure and work up to direct contact. Keeping a distance, walk your dog or puppy on a leash around a chicken yard, coop, run, or enclosure. Watch to see how your dog reacts when he spots the birds. Do his ears perk up? Do his eyes lock onto the birds and become fixated? Or, does he glance their way, sniff the air, and then move on to other smells? Read your dog's body language as you walk around.

If the dog shows signs of wanting to chase the birds (barking, growling, tugging at the leash, all while keeping eyes on the birds), you may want to call it quits for the day and try again another time. A dog that is aggressively interested in chickens may warrant some professional training to become acquainted with and trustworthy around a flock. If the dog is interested but not aggressively so (watching, sniffing the air, but generally moving about) or completely ignores the birds, you can move closer to where the flock is. Let your dog or puppy sniff the chickens and the ground around the enclosure. The nose is the information gatherer of the dog's senses so let your pup sniff for as long as he wants.

Continue to introduce the dogs and chickens in this manner—with your supervision at all times—as you gradually eliminate barriers between the two species. This process may take days or weeks (or more). Be patient and take your time; continue to read your dog's body language and follow his cues.

Can't We All Just Get Along?

There may be some dogs that you feel will never be trustworthy around chickens. This can be an unfortunate reality that you will have to reconcile as a dog owner and/or chicken owner. If you have a dog that continually shows aggression, extreme herding behavior, or relentless chasing, it is likely that this dog's prey drive is very high and may not be trusted loose around a flock of chickens. While not ideal for you, do remember that this is not the dog's fault—it is his instinct and simply in his nature. In a case such as this, it is your responsibility to keep both species safe. First and foremost, train your dog to listen to your basic commands; keep him on leash at all times around the flock. House your flock in a predator-safe coop and lock them up each night. Watch diligently if your dog and chickens are ever in the same area together; and if you can't be there to monitor interactions, keep them separate. If you are unable to take these steps, it would be wise not to keep dogs and chickens together at all.

Teach an Old Dog New Tricks

There are some that believe with enough time, effort, and diligent training, any dog of any breed can learn to live peacefully with chickens. Whether your dog requires daily obedience training or naturally has a nonchalant attitude around fowl, take the time to learn some basic commands and practice these with your dog regularly. Though it may be entertaining to watch, don't encourage play between your dogs and birds; play may be a precursor to more aggressive behavior. Look on the Internet or contact your dog's breed club for qualified trainers in your area that specialize in that breed.

Above all, use good judgment. Read your dog's body language and know the animal well enough to determine which precautionary steps you need to take to keep both dog and chickens safe.

The Guardians

Dogs have guarded livestock for centuries. In my flock, we've tried a variety of roosters to protect our hens from foxes and raptors, but none has been as successful as the simple presence of our large retriever mix. She's not a traditional guard dog: She's a big, playful and goofy mutt more interested in chicken poop than anything else in the yard. But she has a very low prey drive and was raised around our flock from puppyhood, so everyone has had years to become acquainted. She doesn't know she's guarding the flock, but the results speak for themselves. We wouldn't trade her for the world.

If you're looking for a canine protector, here are some great livestock guardian breeds:

- Akbash
- Anatolian shepherd
- Great Pyrenees
- Komondor
- Kuvasz
- Maremma

Komondors, with their unique moplike coats, are best known for being extremely protective guardians.

Four-Legged Flock Members: Other Pets

Another consideration for the pre-chicken keeper is other pets. If you already share your space with and care for other animals, you'll need to consider how well they will fare with chickens and how well chickens will fare around them. Indoor birds, rabbits, guinea pigs, hamsters, gerbils, amphibians, and other similar critters are usually quite compatible with a flock of chickens—mostly because they never cross paths. Other barnyard animals, such as horses, goats, pigs, and cows, were practically made to cohabitate—whole books have been written on successful pasture rotation including ruminants and chickens, but I won't cover that in this book. Other poultry, such as turkey and ducks, on the other hand, can share some nasty parasites with chickens, putting the health of all at risk, so do your research before integrating flocks of poultry.

Generally, cats do not pose a viable threat to most adult chickens. Outdoor cats may cautiously interact with free-range chickens, but there are rarely major conflicts. On the other hand, cats do pose a threat to small chicks and even some fully-grown bantams. Even a seemingly timid house cat can cause serious harm or death to either. Chicks brooded indoors around house cats should be properly protected: Predator-proof the brooder so that cats can't jump into it or reach in and claw at the chicks. If at all possible, brood chicks in a separate room where you can close the door and keep cats away completely. Introductions between the species can happen with your supervision, but the two should never be left together unattended.

While cats have a fairly predictable set of hunting behaviors, dogs are an entirely different matter. Some dogs are very aggressive toward animals they perceive as prey and wouldn't hesitate to kill a chicken (or an entire flock). Others may be disinterested or even compelled to protect the flock. Chickens and dogs may not seem to be the most compatible of species, but with an early start, some good training, and a watchful eye, certain dogs (of any size) can find a very harmonious existence with a flock of backyard chickens. You may find that keeping one of these compatible canines nearby can ward off a number of unpleasant predators all while fulfilling your pup's need for a job.

Home on the Range: Preparing Your Backyard

For just a minute, put this book down, head over to the window, and take a look at your backyard. What do you see? Do you have an expansive green lawn? Flower gardens? Large established trees? How about a deck or patio? Once chickens are introduced into your outdoor space, they will quickly make themselves at home in every nook and cranny within their reach: Flower pots, water fountains, veggie beds—it's all fair game. Through the investigative powers of their beaks, talons, and sheer curiosity, no stone will be left unturned (so to speak).

To immaculately manicured flower beds and perfectly mulched gardens, a flock of chickens can be a frustrating nuisance at best and a devastating force at worst. Chickens scratch at soil, dig craters in the

Tales from the Coop

It's probably becoming clear that I'm quite an animal person. When my husband and I started our first flock, we had three dogs: two Chihuahuas and a shepherd-retriever mix, all rescue animals. Our most timid Chihuahua, Lucy, steered clear of the chickens completely; our boisterous and outgoing Chihuahua, Ruby, happily followed our hens (to eat their droppings, of course) and taunted our rooster by zipping and jumping around him. Whether she was genuinely trying to play with him or gleaning joy simply from irritating him was anyone's guess.

We got incredibly lucky when we adopted Winnie, the shepherd-retriever mix. She was just a few months old when she met the chickens and, since they were pullets themselves, both species grew up together. Once fully grown, Winnie ended up reaching a whopping 90 lb. (40 kg). Lucky for us, she was respectful and gentle around the chickens. She acted as a wonderful guardian when the chickens were free ranging, and, aside from her penchant for chicken droppings, she completely left them alone.

We also fostered dogs for our local shelters over the years, but not all were as respectful of the chickens as our three. Some of the adult dogs we brought home had very strong prey drives. Others were rowdy puppies that were driven solely by the desire to play. Thankfully, no chickens have ever been harmed in our fostering adventures (which continue to this day), but we always played it safe and exercised extreme caution.

dirt for dust bathing, and eat the foliage and fruit from favorite flora. If they find something particularly tasty in your garden, they'll eat it down to the roots. They'll also overturn compost piles searching for bugs and grubs (it is up to you whether that's a boon or a nuisance), without concern for making a mess. In the process, they'll deposit droppings in their wake. With free-range chickens, these scenarios are not a matter of *if* but rather *when*.

Like any other preparation, chicken proofing a garden is a project best started *before* any peeps come home. The easiest and only foolproof way to keep chickens out of areas where they're not welcome is to put a barrier between them.

Fencing

Fencing used to keep chickens out of gardens need not be fancy or even heavy duty for that matter. It only needs to serve as a barrier to keep marauding chickens out (or in, depending on how you use it). This means most of your options won't be terribly costly or difficult to install.

If your chicken fencing serves dual purposes of protecting your gardens from chickens and protecting your chickens from predators, turn to chapter 10 for a more detailed guide on predator proofing. If your neighborhood is home to loose-roaming dogs, it would be wise to consider dual-purpose fencing.

A Great Pyrenees dog makes an excellent livestock guardian..

Assuming you simply want to deter your chickens from destroying your gardens, there are several simple fencing options.

Chain Link. Chain-link fencing is very durable and a common fencing option for many homeowners; however, it is quite expensive to use for simply keeping chickens out of unwanted areas. Consider chain-link fencing for perimeter use or for your own use. If you already have chain-link fencing erected on your property, it will work very well at keeping chickens corralled. Fortunately, for those putting up new fencing, there are many other chicken-friendly options.

Chicken Wire. Though the mesh is too thin for all-purpose predator protection, chicken wire fencing is one of the best options for keeping chickens out of unwanted areas. It's inexpensive (comparable in cost to plastic), supereasy to find, and installation takes minutes. Farm supply stores will often sell metal posts for bracing that do not require hardware or the use of tools; wood posts also work well but will require some minor hardware to install properly. Chicken wire is available in a variety of lengths and widths, is easy to move, and looks clean and rustic in a gardenscape.

Decorative Garden Fencing. Decorative landscape fencing is usually available at home and garden stores and is manufactured in a dizzying array of styles, colors, height, designs, and patterns. Decorative fencing is commonly made from new or recycled plastic or metal. While inexpensive and pretty, decorative garden fencing is usually rather ineffective: Many styles are too low and chickens will simply hop over them. Others are tall but have wide openings through which a chicken can easily squeeze. Decorative fencing may work to deter new chickens for a little while, but they'll eventually become curious and determined to investigate what lies on the other side.

Electric Fencing/Netting. Electric poultry netting has many pros: It's easy to move and set up anew, it's affordable, and it's incredibly effective. Unlike other garden fencing options, electric netting will successfully keep predators out and chickens in. This is a favorite option for many chicken keepers, since it offers the peace of mind of knowing the chickens are safely enclosed. Of course, any electric fencing relies on electricity, so it is susceptible to power outages and other related snafus. If you choose electric netting, keep the weeds and grass mowed below the fencing, since overgrowth can cause the fence to short out. Also investigate solar-powered fencing to keep electric bills down and to have a reliable source of power. Electric fencing isn't as aesthetically pleasing as some other options, but it does the job well.

Tales from the Coop

A life with animals was one I had long known well. The responsibility of their care came naturally and always held joy for me. For the most part, the change in routine that came with keeping chickens felt seamless for my husband and me, city slickers though we used to be. Overall, they fit right into our homebody lifestyle.

But mistakes are made, and nobody is perfect. The first time we experienced a fox attack was early one fall morning with our first flock of hens. I had failed to close up the coop the night before (we had been out to dinner), and the chickens were out roaming in the dewy, unmowed grass (a chore that my husband hadn't gotten to the weekend before). Between the two of us, we had unintentionally set up the perfect conditions for a fox to have a chicken breakfast. By the time I realized what was happening, it was too late for Rita, our white Easter Egger hen. I caught a glimpse of the fox at the edge of our wooded property just as she was fleeing the scene. The vixen paused and turned to look at me, with Rita in her mouth. Then she gracefully dashed into the tree line.

It's important to be as prepared as possible for the lifestyle changes that come with keeping chickens, but it's equally important to forgive yourself when you do your best and accidents happen. Every chicken keeper will experience the loss of his or her birds sooner or later. We can then learn from our mistakes and become better wardens of our birds.

Chain link fencing is a great option to keep chickens corralled.

Hog Fencing. Another handsome fencing option is hog fencing. With its large, square openings (about 4 in. [10 cm] per side), small predators such as weasels, snakes, mice, and rats, as well as tiny chicks and very young birds, may walk through freely, but it is suitable for keeping adult chickens confined to an area. Hog fencing panels are more expensive than either plastic or chicken wire and not very easy to move once installed, but they work well and make great garden fencing.

Plastic. Plastic fencing is popular for its accessibility, its ease of installation, and its low price tag. Made of vinyl or PVC, plastic fencing is available in a variety of mesh sizes, shapes, and colors and makes great temporary fencing. Unfortunately, plastic has several drawbacks: The material is incredibly flimsy and becomes brittle with the fluctuating temperatures of many seasons. Plus, any determined predator can knock it down or chew right through it, getting to your chickens quickly. What's more, broken fencing leads to a lot of wasted material and money.

Wood. A great natural source of fencing material, wood may look beautiful in a garden. Unfortunately, wood privacy fencing is rather large and may require professional installation, adding a significant increase to the cost. Picket fences may be used to partition off gardens but are not easy to move once installed. Depending on the height, chickens may fly up to perch on wood fencing and fly onto the other side, gaining access to everything that is off limits.

Hardscapes: Concrete, Stone, Decking, and Gravel

Patios, decks, stone pathways, and other garden features beautify spaces and make outdoor living easy and comfortable. The chickens that share these spaces will navigate them in much the way humans do: They'll come and go as they please, walking on any surface you or I might walk on. Hardscapes are rarely a deterrent for chicken traffic.

The major considerations when introducing chickens to hardscapes is the challenge of keeping them clean. As chickens graze and meander throughout the day, they'll leave droppings in their wake. You'll likely want to keep heavily trafficked outdoor areas clean to avoid tracking manure into your home or car. If you keep a small flock of hens, cleaning droppings with a cat litter scooper is a viable option but one that will probably become tiresome and tedious over the years. Instead, you may want to install only garden features with excellent drainage.

Deck railings, fence posts, and low fencing all make enticing perches for backyard chickens. As you'll learn in later chapters, the preening and grooming that follows dust-bathing is a social activity

Chickens will find a vegetable garden an attractive source of bugs, worms, seeds, and nuts, so installing a barrier is the only way to protect your valuable plantings.

Know the Cold, Hard Facts

Still wondering exactly what you're getting into? Here are some cold, hard facts to think about before getting started with chickens.

Expect to get dirty. Chickens are messy. They eat like teenage boys, stand on every available surface, and they poop whenever (and wherever) the mood strikes.

Expect to give tours of your coop. For neighbors, friends, and family, your hen's house will become a small attraction (as if you needed another reason to keep it clean).

Expect to become an expert at making scrambled eggs.

Expect to commit. Most battery-cage hens live for only a year and are then pulled out of production and "processed," but a well-cared-for domestic chicken may live up to ten years. While the first two to four years of life are the most productive egg-laying years, most layer breeds will dependably lay for up to seven. If you raise healthy, laying breeds, you'll be pulling eggs out of the nest box for many years.

Expect to make some modifications to your lifestyle. Whether it's sharing a backyard with your feathered friends or traveling a bit less, bringing home any new pet warrants some compromises to your current routine.

Expect that your chickens will take a break from egg laying every once in a while for their health and well-being. A hen's reproductive system waxes and wanes with the seasons. Egg laying naturally declines in winter as the days grow shorter. Other annual events, such as molting, will also put the brakes on laying for a little while.

Expect to be thoroughly entertained. Chickens are, in a word, ridiculous. They have a waddle that's nothing short of slapstick comedy, they make the funniest chuckles and chortles you've ever heard, and their antics never cease. You will never be bored keeping chickens. Promise.

that often takes place in small groups while perching. And as chickens perch, they poop. You can either restrict your flock to certain areas of your backyard or make your peace with poop.

Softscapes: Mulch, Soil, Lawn, and Gardens

Of all the "'scapes" in the garden, softscapes are the most susceptible to destruction by chickens. Soil harbors all manner of grubby bugs, wriggly worms, seeds, and nuts, and seeking out and finding this forage is what chickens live for.

To the gardener, mulch is a must. It protects the precious topsoil, helps the ground retain moisture, and keeps weeds to a minimum. To the chicken, mulch is a flimsy barrier between her beak and the good stuff. In a few swift digs of her talons, she'll send the mulch flying, revealing the vulnerable soil and bugs below. Like a moth to a flame,

chickens instinctively know to search under mulch, piles of leaves, and similar ground covering to find protein-rich forage.

Like mulched areas, flower beds and edible gardens are prone to damage from free-roaming chickens. Vegetable gardens and fruit trees are especially targeted: The sweet greens, juicy fruits, plump veggies, and tender seedlings of newly planted beds are simply irresistible to chickens. With both mulched and cultivated gardens, fencing is the only surefire way to keep chickens from getting to these valuable crops. Fencing can be utilized to corral chickens to one area, enclose entire gardens, or even cover individual beds, so do what works best for your setup and budget.

Generally speaking, grass-covered lawn is less vulnerable than the types of gardens described above. While chickens will dig and scratch at grass, they'll only do significant damage if they are confined to a small patch for a very long time. Chickens will nibble the tips from grass and take bites of weed and plant leaves but will rarely pull up any of these from the roots while grazing. Chickens may be rotated on grass with portable fencing or pens very successfully. (Chicken "tractors" and "arcs" serve much the same purposes; see chapter 6 for more on portable housing.) The size of the pen and the number of birds in your flock will give you an idea of how quickly they'll eat the greens down to dirt. For the first few rotations, you will need to watch them closely and move them before you see significant damage to your grass.

"Misc"-Scapes: Outdoor Furniture, Birdbaths, and Garden Accessories

As with deck railings and fences, some garden accessories, such as outdoor furniture, birdbaths, outdoor grills, and garden art, will be used as perching and preening posts. Remove or relocate pieces that you don't want soiled, especially anything that children play with, such as swing sets or outdoor play equipment, and anything that comes into contact with food, such as grills and tabletops. Furniture that remains outdoors will almost certainly require regular cleaning after perching and preening activities, or you may opt to store it when not in use.

But don't toss the plastic pink flamingos just yet: Many garden accessories are simply of no interest to chickens. My flock shares the garden with a life-sized concrete chicken statue that goes largely unnoticed. New birds will often eyeball the statue with a sidelong glance, but once they determine that the gray hen is not a threat, they'll go about the more important business of tearing up my flower garden in search of crickets.

None of this information is meant to discourage you from allowing your chickens to free-range. In fact, foraging—and all of the great health benefits that come with it—is an essential aspect of any chicken's life. As with any other area of chicken keeping, I listed these concerns because a few preventive steps can go a long way toward derailing any unfortunate surprises for a new chicken keeper. So, find a happy medium: Protect your gardens, but let your chickens have some free-range time. It's entirely possible to have a beautiful, flourishing garden, clean outdoor spaces, and happy, free-roaming chickens.

Chapter 3

The Ins and Outs of Chickens

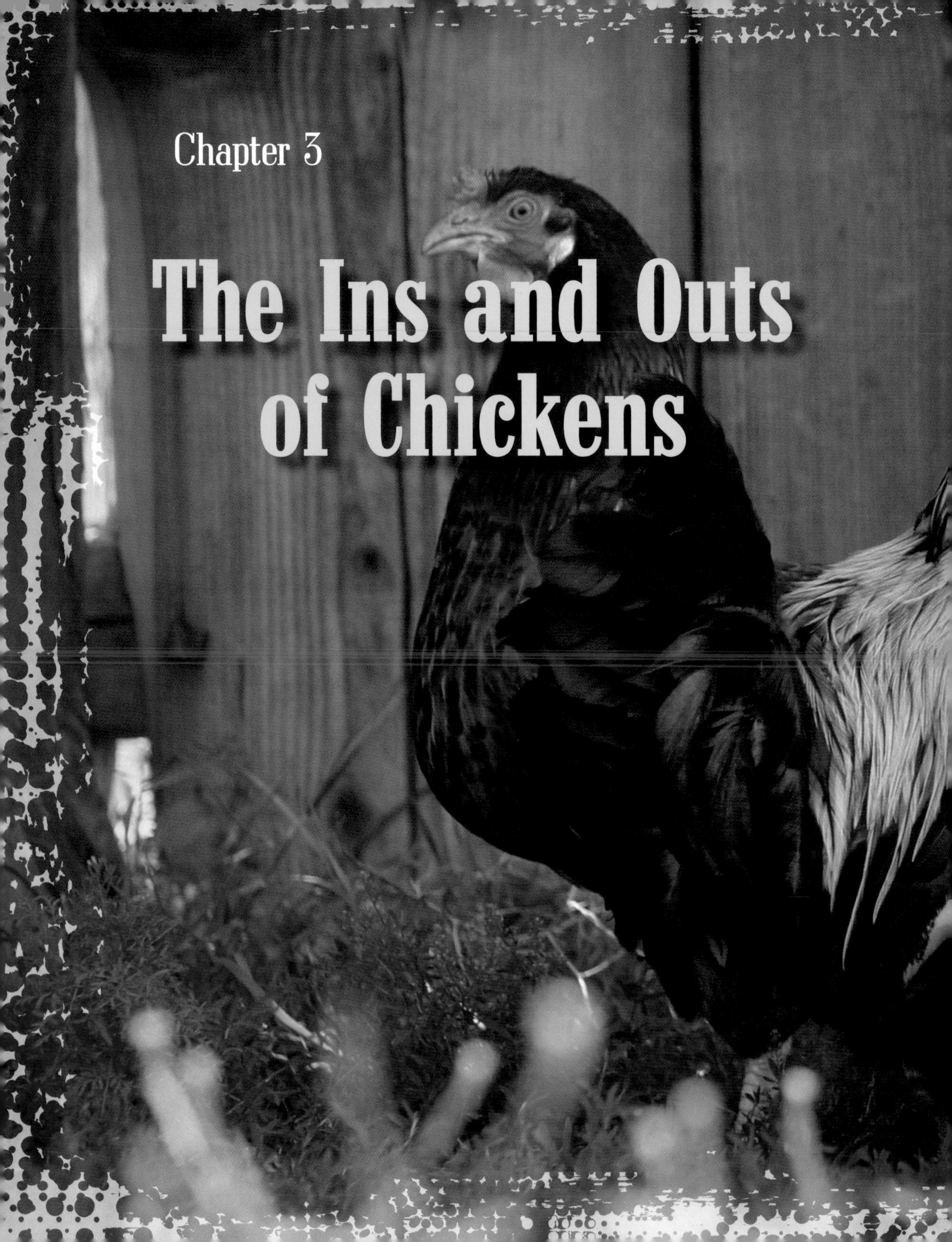

With so many breeds and varieties of chickens to choose from—not to mention the ability to order online with a few clicks and have them in your hands weeks later, you'll have some initial choices to make before bringing home the babes. Your choices will range from the 4 lb. (2 kg) game bantam to the 10 lb. (5 kg) Jersey Giant, and between those two extremes are birds with frizzled feathers, beards and muffs, and genes that allow them to lay green and blue eggs. Some breeds are silly and sociable, and some can be standoffish and aloof. Some are flighty, and some are bold. While the topic of chicken breeds could easily fill a whole book (and it has), this chapter will focus on the breeds that are best for the suburban backyard. These birds are primarily egg-laying champions and superfriendly birds. Most of them forage well or live in the smaller coops afforded to urban or suburban backyards, and they're all really good looking to boot.

But before we get to the *outs*, let's start with the *ins*. Large or small, there's one thing that all of these breeds have in common: their anatomy. While knowing the ins and outs of chicken behavior will help you provide them with all they need to be comfortable and content, knowing the ins and outs of chicken anatomy will help you catch early signs of disease, symptoms of infestations, and will, overall, turn you into a better chicken keeper.

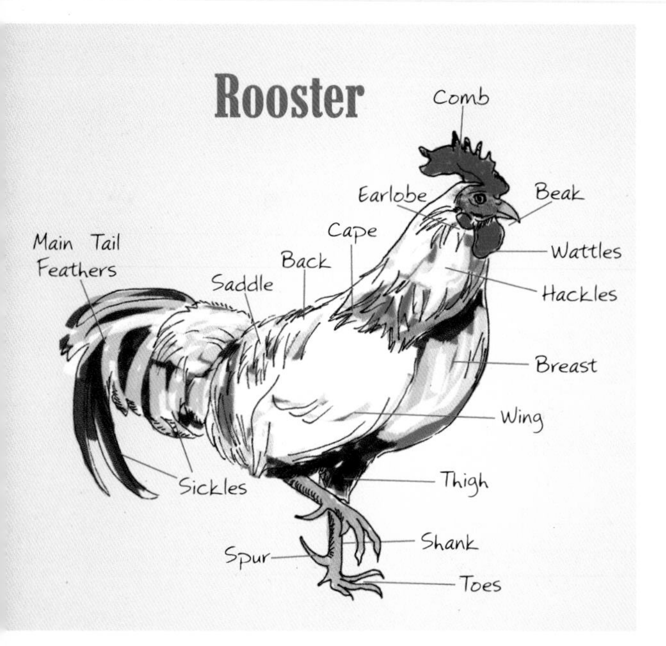

Chicken Anatomy 101

Chickens may seem like simple creatures to the naked eye, but nature has expertly crafted them to thrive with scarce food sources, find an appropriate mate, safely produce and protect their young, and so much more. Knowing what makes them tick is part of the joy of keeping chickens.

Our many varieties of chickens were cultivated and bred from a humble ground fowl native to Southeast Asia. It is thought that both the red jungle fowl and gray jungle fowl provided some genes to the domesticated chicken we know today, and as such, our various modern-day birds have a few things in common to all.

For starters, all domestic chickens have two legs, two wings, and feathers. (I know, tough stuff, here. Are you taking notes?) With the exception of the Araucana breed (see page 50 for more on them), all chickens have a tail. Most have four toes, but a few breeds have five. And regardless of gender, all chickens have a comb.

The Outs

The **comb** is a fleshy, featherless patch of skin on the top of a chicken's head. Both male and female at all life stages have one, but an adult male's comb is usually the most pronounced. Combs range in color from bright red to dark maroon and reach their deepest color around sexual maturity. The styles of combs include the single, rose, pea, buttercup, cushion, and strawberry, among others. The males' large comb is helpful when attracting a mate, but the comb's main function is to regulate the body's temperature through blood circulation. Breeds suitable for warmer temperatures tend to sport a larger comb (to release more heat), and those bred for colder climes have a smaller, more compact comb.

The **eyes and ears** are located on the chicken's head. Like most birds, chickens are sensitive to light and can distinguish colors. Exposure to light triggers hens to lay, and using a colored lightbulb (red is preferred) when brooding chicks can reduce pecking and fatigue.

To find a chicken's ears, you'll have to know where to look. The ear is a tiny opening on either side of the head surrounded by a fleshy earlobe distinguished by color. Generally, a hen with a red lobe will lay brown eggs, and a hen with a white lobe will lay white eggs, although there are a few exceptions.

The **wattles** are peculiar indeed. These two featherless flaps of skin hang from the bottom of the face and can range in color from red to blue to black. Like most parts of chicken anatomy, they are usually far more pronounced in roosters.

Common Combs

Single

Buttercup

Rose

Pea

The **body** of a chicken is generally shaped like a horseshoe with a high head and tail. The *breast* is located below the head and neck, above the belly. The *saddle* is the area of feathers between the neck and the tail and is often quite colorful in males. *Tails* vary tremendously in color, arch, and size, especially between the genders. A rooster can often be distinguished by his brightly colored and high-arching tail feathers.

The **legs and feet** of a chicken may be bright to pale yellow, white, slate, olive, or black, all depending on breed, of course. The feet have either four or five toes (unless the bird was very unlucky indeed), and the legs are covered in overlapping slabs of skin called *scales*.

Spurs are sharp, bony protrusions on the back of the leg used for fighting and for defense by the valiant rooster. Contrary to popular belief, hens also have spurs, although they are usually quite small.

The **skin** of a chicken, which also varies in color depending on breed, is thin and tears easily. Don't be fooled by feather color, since it does not necessarily indicate the bird's skin color. Most have white or yellow skin, with the exception of the Silkie's black skin. Like the yolk of a hen's egg, a chicken's skin will deepen in color if she is pasture raised and eats more bugs, greens, seeds, and grasses.

A chicken's feathers are one way to identify her breed; they also indicate her overall health and vigor. A chicken's feathers should be glossy, smooth, and bright.

The **feathers** are an easy way to distinguish some of the breeds, but many chickens share similar colors and patterns. Feathers cover most of a chicken's body, and in some breeds, such as the Brahma and the Faverolles, the birds have feathers on their legs or feet.

The variations in feather pattern are quite plentiful, too. In addition to a variety of earth-toned colors, feathers can have *striping, penciling, wide* or *narrow lacing, barring, spangling,* or *stippling.* Despite their alluring colors and styles, chickens lose their feathers each fall or winter in a process called *molting.*

Feathers don't stop at the neck, either. Facial feathers include *muffs,* which are furnishings located on the side of the face, common in several breeds. A *beard* is a grouping of feathers below a chicken's beak, and *tufts* are unique to the Araucana chicken, characterized by feathers growing from a small space near the ear lobe. Some breeds, such as the Polish, sport *top knots:* groupings of feathers in a pom-pom–like shape on the top of the head.

The key to feather health is allowing your birds to dust-bathe and preen as they see fit. (See chapter 8 for more on dust-bathing, preening, and molting.)

The Ins

The **digestive system** begins with the beak. As a bird collects forage or feed in her beak, she adds

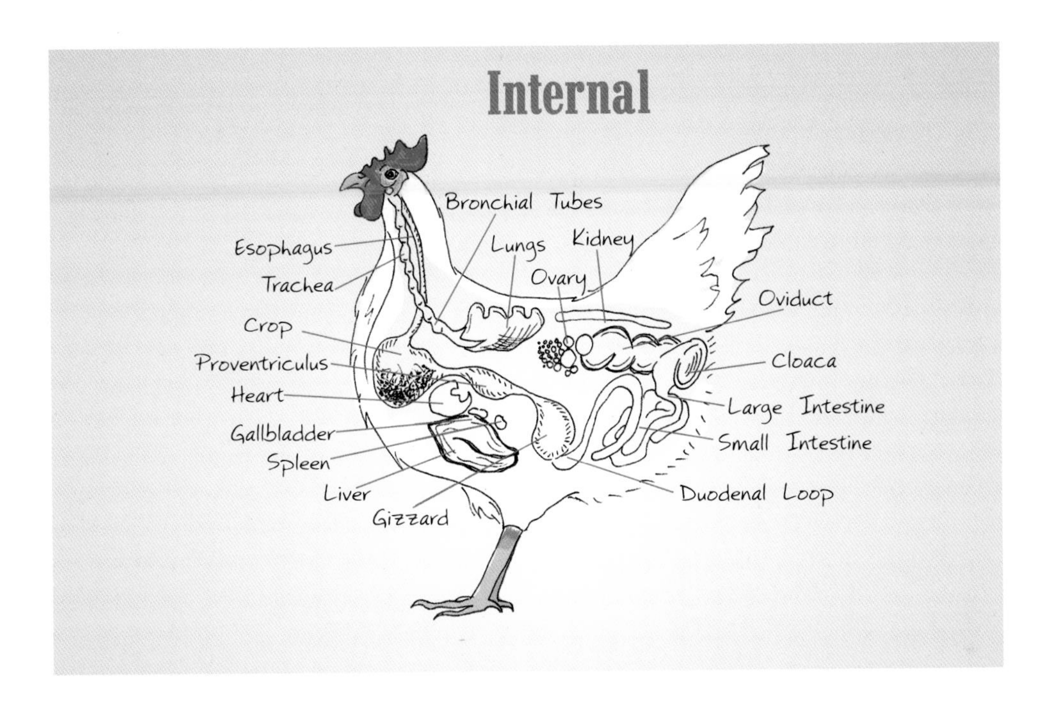

some necessary digestive enzymes (contained in saliva) before the food, usually still whole, passes through her *esophagus*. Chickens do not have teeth, so the "chewing" process happens later (giving credence to the adage "as scarce as hen's teeth").

The *crop*, a flexible pouch located just below the chicken's neck, is the next stop. When a bird is finding plentiful forage, the crop will swell considerably and house all of the bird's food, from grasses to grubs. The more difficult to digest forage may remain in the crop upward of 12 hours. Consider the crop a holding area for the bulk of a bird's sustenance.

Next, the food is moved to the bird's true stomach, or *proventriculus*, where more digestive enzymes are added. The *gizzard* is next, where the grit you've provided your birds (or the pebbles they've picked up foraging) is put to use. Without any teeth, chickens rely on the teamwork of grit and the strong muscles of the gizzard to break down their food. After absorption of the nutrients by the *intestinal walls*, liquid and solid waste is released together through the bird's *cloaca*, and digestion is complete. Droppings are a key indicator of health, so monitor them from time to time. The white crest on a normal, healthy dropping is the urine deposit; the rest should be light greenish brown or gray. It's not pleasant, but don't be too grossed out—that "waste" is compost gold, so put those nutrients to work a second time in your garden.

The **reproductive system** is the system most chicken keepers are concerned with; after all, eggs are the reason most people start keeping chickens. Knowing a bit more about what it takes to produce that egg will make you a better keeper.

The cock's reproductive role is centered mostly on two *testes*, which unlike male mammals, are housed internally. In a flock of mixed genders, mating occurs continuously. For hens, reproduction begins in her *ovary*, where she develops small clusters of *yolks*.

Tales from the Coop

When my husband, daughter, and I moved into our home in North Carolina, we "inherited" a resident chicken—a single, sweet, skittish Delaware we named Hen Solo. She was the sole survivor of her flock, all of which had been picked off by predators over the years. By the time we started caring for her, she had been accustomed to roaming the homestead freely, eating only what she could forage (including water) and having the coop to herself.

We immediately put her on our flock's diet of organic, soy-free feed, with grit and oyster shells as supplements. She had fresh water daily. She wasn't laying eggs when we moved in, and she continued her dry spell through the winter and into the spring. When the days lengthened, she surprised us with one perfect beige egg. We were ecstatic. She was isolated from the rest of our flock at the time (for biosecurity reasons), so we knew the egg was hers. She was back in business.

Several weeks later, my husband came home from work one evening and collected eggs on his way in. He reported that there had been three eggs in Hen Solo's nest box and wondered why I hadn't collected eggs in a few days. I looked at the three perfect beige eggs in his hands in wonder: The truth was, I had collected eggs the night before. Everything I knew told me that it was impossible for a hen to lay more than one egg per day. But the proof was right in front of me. There was no doubt about it: Hen Solo had laid three eggs in one day!

Through *ovulation*, yolks are released into the *oviduct*, a 25 in. (64 cm) long tube. If the hen has been mated, fertilization will happen there. (And if not, the process will still continue; you just won't get a fertilized egg out of the deal.) The yolk moves through two parts of the oviduct, the *infundibulum* and the *magnum;* the latter is where the *albumen* (or egg white) is added. The third part of the oviduct, called the *isthmus,* is where the hen's system adds the *shell membranes*. Finally, a large deposit of calcium carbonate (mostly pulled from the hen's bones) produces the shell in the *uterus* (this is why it's so important to provide added calcium to your laying hens in the form of oyster shells). Finally, after a 20-hour process, the internal chamber called the *cloaca* releases the fully formed egg through the *vent,* encased in *bloom*. Also called the *cuticle,* the bloom is a protective coating that safeguards the developing chick from bacteria that may penetrate the shell.

Pretty cool, right? Not only is the egg a perfect food, it's a perfect structure to incubate the young bird. Most young female chickens, called *pullets,* reach sexual maturity between four and six months, depending on the breed and the individual bird's lineage. When the days lengthen and the pullet is exposed to over 14 hours of daylight, her reproductive system kicks into high gear and starts egg production. Most high-production breeds will lay one egg per 24- to 26-hour cycle. If it takes longer to acquire those 14 hours of light, egg production will carry over into another day or two. Some chicken keepers choose to artificially light the coop to encourage their chickens to keep laying through the shorter days of winter. This is a widely accepted practice, but I think the winter months are a well-earned break for everyone. Considering that shortened daylight hours usually coincide with a molt, the fall and early winter are the perfect time to let the birds rejuvenate and rest. Plus, some breeds make great winter layers, no matter what you do.

Choosing the Right Breed

The breed or breeds you choose to raise will largely be determined by two factors: your region's greater climate and the space you have available to devote to your flock. Once those two considerations are met, decide how important other factors are, such as egg production, personality, and temperament. The primary purpose of your flock will determine which breed(s) you'll raise most successfully.

There are hundreds of chicken varieties, each with its own strengths. The descriptions that begin on page 44 highlight some of the most popular American chicken breeds for the backyard flock. I'll admit, I'm biased: Many of these are my favorite chickens, and I wouldn't have a flock without them. Just keep in mind that while many generalizations can be made for a breed as a whole, individual birds may not always conform to their breed's exact temperament, size, color, or other factors. Many of the following chickens are heritage breeds, dual-purpose (which means they can be used as both egg layers and meat birds), and some fall on The Livestock Conservancy's (TLC) Conservation Priority List for endangered poultry (see below).

Lineage aside, what these breeds all have in common is that they're easily handled, easy to find and purchase, friendly toward their keepers, and laying eggs is their strong suit, making them the absolute best birds for the suburbs.

The Livestock Conservancy

The Livestock Conservancy does research, education, outreach, marketing and promotion, and genetic rescues to help ensure the future of rare breed agriculture.

Each year, The Livestock Conservancy publishes its annual Conservation Priority List for endangered poultry. The breeds identified on the list generally conform to certain parameters, as identified below.

Critical. Fewer than 200 annual registrations in the United States and estimated global population less than 2,000.

Threatened. Fewer than 1,000 annual registrations in the United States and estimated global population less than 5,000.

Watch. Fewer than 2,500 annual registrations in the United States and estimated global population less than 10,000. Also included are breeds that present genetic or numerical concerns or have a limited geographic distribution.

Recovering. Breeds that were once listed in another category and have exceeded "Watch" category numbers but are still in need of monitoring.

Study. Breeds that are of genetic interest but either lack definition or lack genetic or historical documentation.

Source: The Livestock Conservancy at www.livestockconservancy.org

The Best Egg-Producing Breeds

Ameraucana

Description: The Ameraucana is best known for its ability to lay eggs in various (and lovely) hues of blue. These eggs are a stunning addition to any egg basket or market table, and they make a great show-and-tell project for school. In the looks department, Ameraucanas themselves are pretty cute, too—they sport facial furnishings and are available in a wide variety of colors and feather patterns. They have winning personalities and are fun birds to have in any flock.
Productivity: Good to very good
Egg Color: Blue
Temperament: Calm, non-aggressive, friendly, easily handled
Hardiness: Cold-hardy
Varieties: Black, blue, blue wheaten, brown-red, buff, silver, wheaten, white

Australorp

Description: Of all the heritage chicken breeds, the Australorp is widely considered the best for egg production. The Australorp has its origins in Australian-bred Orpington chickens, hence the name (Australia + Orpington = Australorp). The Australorp is a medium-sized bird, and hens are prolific layers of large, brown eggs. The breed's one color, black, is a stunning display of subtlety, with hints of green and purple in the right light. They are really gorgeous birds. Overall, Australorps are friendly and personable in temperament, although some individuals may have the tendency to be the enforcers of hierarchy in the flock. They are very friendly and curious toward their keepers, though. If you're looking for the best heritage breed layer, this is it.
Productivity: Excellent
Egg Color: Brown
Temperament: Docile, easily handled, quiet disposition; may dominate other birds
Hardiness: Cold-hardy
Varieties: Black

Brahma

Description: The Brahma is a gentle giant. Quiet, calm, and even-tempered, this breed is exceptionally cold-hardy, and the hens are great winter layers. Because of their feathered legs and shanks, they're not best for excessively wet or muddy regions (moisture can cling to the feathers, leading to frostbite on toes). Otherwise, the Brahma is a great addition to the backyard flock, especially in northern regions. Despite their large size, they handle confinement very well and are not quite as active as other breeds that prefer to forage; this makes them great for smaller spaces, too. The breed's easy-going temperament makes it a perfect chicken for kids.
Productivity: Good
Egg Color: Brown
Temperament: Very calm, quiet, non-aggressive, easily handled, very gentle
Hardiness: Cold-hardy, not heat-tolerant
Varieties: Light, dark, buff

Ameraucana

Black Australorp

Brahma

Buckeye

Description: The Buckeye chicken gets its name from its state of origin, the Buckeye State of Ohio, where Nettie Metcalf—the only woman to develop a recognized American breed of chickens (so far)—cultivated this deeply hued chestnut bird. Although the Buckeye and the Rhode Island Red originated around the same time and were recognized as breeds in the American Poultry Association's (APA) Standard of Perfection in the same year, they should not be confused with one another; they are totally separate breeds. Robust and meaty, the Buckeye is active, preferring to forage for food and likes to have some room to roam but can adapt to confinement. Hens lay a medium-sized brown egg, and the breed is dual-purpose. The Buckeye is currently categorized as "Threatened" on The Livestock Conservancy's Priority List.
Productivity: Good
Egg Color: Brown
Temperament: Very friendly, very active, great forager
Hardiness: Cold-hardy
Varieties: Dark reddish brown only

Delaware

Description: This lovely bird is relatively new in the world of chicken breeds, developed in 1940 in the state that shares its name. The breed is a cross between the Barred Plymouth Rock and the New Hampshire, sporting a unique feather pattern, similar to the Columbian white coloring. Delawares are friendly birds that grow to maturity rather quickly for a heritage breed. Hens are reliable layers, and the breed bears confinement well. The Delaware is currently under "Threatened" status on The Livestock Conservancy's Priority List.
Productivity: Very good
Egg Color: Brown
Temperament: Calm, docile, friendly
Hardiness: Cold-hardy
Varieties: White (similar to Columbian coloring, but with barring where black feathers would be)

Dominique

Description: The Dominique gets the title of America's oldest chicken breed. Easily (and often) confused with the Barred Plymouth Rock (with whom it has a bit of a shared history), the breed experienced a rise and fall in popularity since its establishment in the APA's Standard of Perfection in 1874. Following World War II, the breed nearly went extinct, and only four flocks remained by 1970. Through concerted effort, the breed was revived and now boasts "Watch" status on The Livestock Conservancy's Priority List. If you're interested in raising a fun backyard bird and want to support its ailing numbers, the Dominique is a great choice. They are calm, docile, and the hens are rather nurturing as mothers.
Productivity: Good
Egg Color: Brown
Temperament: Docile, generally calm but can be flighty; hens make excellent mothers
Hardiness:Cold-hardy and heat-tolerant
Varieties: Black-and-white barred only (also known as cuckoo pattern)

Easter Egger

Description: Yes, it's true: The Easter Egger is *not* officially a breed. So how do they make it to the best breeds list? For starters, they are easy to come by from nearly any hatchery, making them quite accessible to chicken keepers in any region. They are really fun to hatch and raise (they don't breed true, so hatching is always a surprise), and like any good mutt, they're superfriendly and full of personality. The Easter Egger trademark is that they are crossed with birds with the blue- and green-egg laying gene, and they lay beautiful eggs in nearly every color. But be warned: Not all Easter Eggers lay blue or green eggs—that's part of the surprise. (See "So What's an Easter Egger?" on page 53 for more on this fun chicken.)
Productivity: Good to very good
Egg Color: Various shades of blue, green, olive, brown, cream, and pink
Temperament: Varies; usually very friendly, docile, easily handled
Hardiness: Cold-hardy
Varieties: Varies

Buckeye

Delaware

Dominique

Easter Egger

Jersey Giant

Description: The Jersey Giant is indeed from New Jersey and indeed giant. Developed with the intention to replace the turkey as a table bird, this dual-purpose chicken takes about eight to nine months to reach maturity (compare that with other meat birds' six *weeks* of growing time). As their name implies, they are the largest chicken breed—roosters weigh in at around 10 lb. (5 kg) and hens around 8 lb. (4 kg). Due to their size, they're not the most economical layer in terms of feed conversion, but they lay better than most large breeds and do bear confinement well, a perk for small backyard coops and spaces. The Giant's easy-going disposition is another reason it makes a great small-flock bird. Black was the original color for Jersey Giants, but they are now recognized in White and Blue as well. This heritage breed is under "Watch" status on The Livestock Conservancy's Priority List.

Productivity: Good to very good

Egg Color: Brown

Temperament: Gentle, mellow, good-natured, easily handled; bears confinement well

Hardiness: Cold-hardy

Varieties: Black, white, blue

New Hampshire

Description: Introduced in 1935, the New Hampshire chicken is a relatively new American breed as well. Deviated from the Rhode Island Red and named for its state of origin (It's just "New Hampshire"; "*New Hampshire Red*" is actually a misnomer), the New Hampshire is a dual-purpose medium-sized bird with a single comb. Its plumage is a coppery red with more gold than Rhode Island Reds. Hens are fairly good layers of medium-sized brown eggs, are prone to go broody, and make excellent mothers. New Hampshires have a "Watch" designation on The Livestock Conservancy's Priority List.

Productivity: Good to very good

Egg Color: Light brown to medium brown

Temperament: Calm, docile, easily handled

Hardiness: Cold-hardy

Varieties: Light brownish red

Orpington

Description: This English dual-purpose breed hails from the town of Orpington in Kent, from which it gets its name. The Orpington is an excellent dual-purpose bird; the hens are first-rate layers of a light brown egg, and both sexes are incredibly cold-hardy thanks to their fluffy, loose feathers (those loose feathers also make the bird look heftier than she actually is). The Orpington breed is recognized in several color varieties, with Buff, a beautiful golden-yellow color, being the most popular. Like the Australorp, this breed has "Recovering" status on The Livestock Conservancy's Priority List. They are a favorite of family flocks and a perfect breed for children to care for due to their gentle nature.

Productivity: Very good to excellent

Egg Color: Very light brown

Temperament: Very calm, docile, sweet, gentle, easily handled

Hardiness: Cold-hardy

Varieties: Buff, black, blue, white

Plymouth Rock

Description: As far as American heritage breeds go, the Plymouth Rock has it all: She is hardy but docile, fares well in colder weather, produces an excellent number of brown eggs, and has a winning personality. Named for the location where the Pilgrims supposedly made landfall, this dual-purpose bird was once the most popular chicken in America. If you're thinking of a bird to set and hatch a clutch of eggs, the Plymouth Rock makes an excellent broody hen and mother. While many colors are accepted for the breed standard, the Barred variety was the original and remains the most recognizable for the breed today. The Plymouth Rock is "Recovering" on The Livestock Conservancy's Priority List.

Productivity: Very good to excellent

Egg Color: Light, with a pink tint, to medium brown

Temperament: Friendly, docile, curious

Hardiness: Cold-hardy

Varieties: Barred, white, buff, silver-penciled, partridge, Columbian, blue

Jersey Giant

New Hampshire

Orpington

Plymouth Rock

Rhode Island Red

Description: The Rhode Island Red is a classic. Perhaps the most popular American chicken and the most well-known domesticated fowl around the world, the Rhode Island Red is an egg-laying rock star. The breed was developed in its namesake state of Rhode Island and is now the official state bird as well. Hens lay a medium-brown egg and are incredibly curious, friendly, and personable. Roosters can become aggressive, however, and are not recommended around children. The common Rhode Island Red available today through most hatcheries is a lighter brownish red than the older, non-production strain, which has "Recovering" status on The Livestock Conservancy's Priority List. Either strain you end up with, this bird is a classic addition to the backyard flock.

Productivity: Excellent

Egg Color: Brown

Temperament: Friendly, curious, sometimes flighty; males can be aggressive

Hardiness: Cold-hardy

Varieties: Red only (Rhode Island white a separate breed)

Sussex

Description: The Sussex is a nearly perfect homesteading chicken. While they are dual-purpose (many heritage breeds are, as you're learning), Sussex hens are reliable layers and have a reputation for laying dependably through winter. As far as housing goes, they're flexible; they bear confinement well (and don't eat as much as some other breeds); or, if left to pasture, they will actively forage for their dinner. They're friendly and very curious by nature. The Speckled variety has a unique feather pattern that effectively camouflages the bird from most predators—a boon when free-ranging. On The Livestock Conservancy's Priority List, the Sussex is also categorized as "Recovering."

Productivity: Good to very good

Egg Color: Cream to light brown

Temperament: Curious, friendly, mellow, active; bears confinement well or actively forages

Hardiness: Very cold hardy

Varieties: Speckled, red, light

Welsumer

Description: The Welsumer is a beautiful breed. Hens sport a partridge feather pattern in a rainbow of browns and golds, and the Welsumer rooster isn't too hard on the eyes either. He was made famous as Cornelius the rooster on the Kellogg's Corn Flakes box in the company's 1950s ad campaign and is usually the bird that springs to mind when you think of a "rooster." Temperamentally, the Welsumer is an intelligent, active, and friendly breed that forages extremely well, offering the chicken keeper an opportunity to save on feed costs. Hens also lay beautiful chocolate brown eggs that are sometimes speckled.

Productivity: Good to very good

Egg Color: Dark brown, sometimes speckled

Temperament: Active, friendly, docile; forages well

Hardiness: Cold-hardy and heat-tolerant

Varieties: Partridge

Wyandotte

Description: The Wyandotte's name is derivative of the Wendat (or Wyandot) Native American tribe of the northeastern region of the United States, although the bird's exact heritage is unknown. The Silver-Laced variety was the only color available when the bird was admitted to the APA's Standard of Perfection in 1883 but is now bred in several striking colors. The Wyandotte is another dual-purpose bird, matures quickly (which means you get eggs sooner), is a reliable layer, and has a great personality to boot. On occasion, an individual may dominate submissive birds in a mixed flock, but they are very cold-hardy and adapt to confinement well. Their striking feather patterns are hard to miss.

Productivity: Very good

Egg Color: Light to brown

Temperament: Easily handled, docile, calm, friendly; may dominate others in a mixed flock

Hardiness: Cold-hardy

Varieties: Silver-laced, golden-laced, partridge, silver-penciled, Columbian, white, buff

Rhode Island Red

Speckled Sussex

Welsumer

Wyandotte

From the Strange to the Bizarre

Some chickens can be really funny-looking creatures. Below are descriptions of three of the strangest breeds.

The Naked Neck

This dual-purpose breed is known for its trademark look—a long neck bare of feathers. The bird's lack of feathers is actually a boon for this old breed: The Naked Neck makes a great warm-weather bird, the carcass is easy to pluck clean (when being raised for meat), and the Naked Neck can very efficiently convert feed to meat and eggs because it uses less protein on feathers compared to other breeds. Naked Neck hens are great layers and excellent mothers, too. The Naked Neck is sometimes called a turken because it looks like a cross between a chicken and a turkey, but let's set the record straight: Contrary to popular myth, the Naked Neck is not a cross between a turkey and a chicken; it's just a funny-looking chicken breed through and through.

Frizzling is a unique dominant trait that results in feathers curling up and out, rather than lying flat against the bird.

As his name implies, this Naked Neck rooster sports a long neck bare of feathers.

The Frizzle

The frizzle isn't a breed unto itself but rather a feather variation available in some breeds (you may see the frizzle trait acknowledged after the breed name, as in Polish frizzle). The frizzling appearance is attributed to feathers that curl up and outward, rather than lying flat against the bird's body. Frizzling is a dominant trait, so it's fairly easy to introduce into your flock through mating and hatching your own birds. Like other ornamental birds with unique plumage, frizzled breeds tend to have less protection from the elements and aren't able to fly (which is a plus if you're trying to keep them fenced in). For these reasons, their keepers need to take additional security measures against cold and damp conditions and predatory animals. Even so, many fanciers feel they're worth the effort: They're quirky, eye-catching, and fun to raise.

Despite their demure size and unusual looks, Silkie females make wonderful broody hens and excellent mothers.

The Polish

The Polish is famous for its crest (also called a topknot), or the poof of feathers adorning its head. While amusing and ornamental, the bird's crest isn't all that practical; it tends to obscure its vision, making it highly susceptible to aerial predators and bullying from other birds in a mixed flock. The Polish recognizes this disadvantage and tends to be a bit skittish and flighty in temperament as a result. However, the hens can be excellent egg layers, and they rarely go broody, making it a great backyard breed. Indeed, a flock of Polish chickens is a sight to behold.

The Polish breed is known for its flashy, if rather inconvenient, crest of feathers.

The Silkie

Silkies are as sweet as they are silly looking. Their fluffy appearance is attributed to their unique feathers; the long feather barbs and thin feather shafts don't "lock" to create a stiff feather like other chicken breeds. Instead, the feathers resemble (and feel much like) the down of a chick, or even fur. Fanciers and backyard keepers alike love the Silkie's ornamental look. The breed is also unique in that it has black skin and bones and five toes instead of the typical four of other breeds. Hens are decent layers of cute, small, cream-colored eggs, frequently go broody, and are widely known for making excellent mothers.

Because of their feathers, the Silkie has a few special management considerations. First, they don't fare well in cold, wet climates; in those regions, Silkies need reliably heated housing and protection from the elements. They're also unable to fly, so they can't perch or escape dangerous situations, making them very susceptible to predators. They're on the small side, so even small dogs and housecats are predators to this breed.

For all of the challenges in keeping Silkies, there are many reasons they're worth the effort. First, they're heat tolerant and are great for warm climates where a lot of the heavier breeds would suffer. They are also very tame birds and are considered the "lap dogs" of the chicken world. Sometimes labeled the Silkie bantam, the breed has been developed specifically to be petite, so it is not a true bantam, but thanks to their demure size, great temperament, and unique appearance, they make a wonderful breed for kids.

Cracking the Case of the Blue-Egg Layers

Ameraucana, Americana, Araucana, or Easter Egger? Which breed are you getting when you order from a hatchery or visit a breeder? What's the difference? Will they all lay blue eggs?

For years, the distinction between the blue-egg–laying breeds has eluded backyard chicken keepers. Myths and misconceptions about the breeds have led to more confusion than ever, from the origin of the blue-egg layer to the cholesterol level in her eggs.

What do you say we finally put those questions to rest?

Blue Layer Origin

Search for reputable breeders when shopping for true Ameraucana and Araucana eggs, chicks, or pullets to add to your flock.

This is about to get a bit academic, so bear with me. The blue-egg layers we know today are descended from two varieties of chickens native to the coast of Chile. They were called the Collonca and the Quetro. Raised by the Araucana Indians of the region, the Collonca variety laid blue eggs, was rumpless, and had a small single comb. The Quetro variety had a tail, was tufted, and laid brown eggs.

Blue-Egg–Layer Breeds

There are three blue-egg–laying breeds, and it can be confusing to distinguish among them. Use the table below to learn some of the differences.

	Name of Breed		
Characteristics	Araucana	Ameraucana	Easter Egger
Comb	Pea	Pea	Any
Muffs	No	Yes	Can
Tail	No (rumpless)	Yes	Yes
Tufts (ear)	Yes	No	Can
Beard	No	Yes	Can
Egg color	Blue or turquoise	Blue	Can carry dominant blue-egg gene, resulting in shades of blue, green, olive, and more
Color/feather variety As recognized by American Poultry Association (APA)	Black, black-breasted red, golden duckwing, silver duckwing, white	Black, blue, blue wheaten, brown-red, buff, silver, wheaten, white	Any
As recognized by American Birding Association (APA)	Black, black-breasted red, blue, buff, silver, white	—	—
Standard or bantam	Both	Both	Both

The birds were popularized in 1921. A chicken expert in Chile, Dr. Ruben Bustos, bred the Collonca and the Quetro, developing a variety that was rumpless, had tufts, and laid blue eggs. He called them *Collonca de Artes*, meaning Collonca with earrings. At the first World Poultry Congress that same year, famed Spanish poultry expert Dr. Salvador Castello reported on his observations of Dr. Bustos's birds, generating much excitement.

Within the decade, the blue-egg layer was imported to the United States from South America. The bird that arrived was not exactly the specimen of a refined breed, however. Better described as an assortment of several different breeds of native chickens, this stock laid the foundation for North American breeders to develop the two blue-egg–laying breeds that we know today: the Araucana (pronounced air-ah-CAW-nah) and the Ameraucana. Since gaining recognition by the American Poultry Association (APA), the organization that provides written descriptions of all the standard breeds of poultry in North America, each is now a breed in its own right. This shared heritage is what causes so much confusion between the breeds today.

Differentiating the Breeds

For many years after the blue-egg–laying chicken's arrival on North American shores, breeders took liberties with the traits they bred into and out of the bird. When the APA chose to create standards for the Araucana breed in 1976, some breeders chose to focus on traits that had been left out of that standard. It caused some upset among breeders who had taken alternate routes with the bird's looks, such as breeding in tails to the otherwise rumpless breed. Those breeders decided to organize and had their breed officially recognized by the APA in 1984. Originally called the American Araucana, the name was eventually condensed to what it is today, the Ameraucana.

Ameraucanas and Araucanas lay beautiful sky blue eggs; Easter Eggers may lay a variety of colored eggs, including blue, green, olive, pink, and brown.

Araucana or Ameraucana?

Both breeds lay blue eggs, but to chicken enthusiasts and breeders, the two are otherwise very different. Because both Ameraucanas and Araucanas must conform to a set of standards to be named as one or the other, the qualifications can be dizzying to many hobbyist backyard chicken keepers (including this one).

For starters, Ameraucanas have full, flowing tails and *muffs*—facial furnishing similar to beards. Ameraucanas do *not* have *tufts*, which are often confused with muffs. Ameraucanas have shanks that are slate to black in color. Though shell color is still a work in progress for Ameraucana breeders, all true Ameraucanas should lay blue-shelled eggs.

Now let's look at the Araucana.

Rumpless and tufted, Araucanas were bred from the imported North American stock to more closely resemble the original Chilean chicken bred by Dr. Bustos. Rumplessness is characterized by a lack of "tailhead," or the long, flowing feathers of the tail common to most chicken breeds. Another unique characteristic of the Araucana is her *tufts*. While *muffs*, mentioned above, are common to other breeds of chickens (such as Silkies, Faverolles, and Houdans), the tuft feather feature, located near the ears, is unique to the Araucana alone.

Both Ameraucanas and Araucanas have pea combs and both lay blue eggs. Both breeds are bred and recognized in standard and bantam (miniature) sizes, and each breed has its own criteria of colors and varieties.

Are you thoroughly confused yet? No? Well, good!

So What's an Easter Egger?

To make matters more complicated, there is a third blue-egg–laying chicken. Often mislabeled as an Araucana or Ameraucana (or misspellings thereof, such as *Americana*), the Easter Egger is not actually a breed. In fact Easter Egger is a catchall name for any mixed chicken that carries the dominant blue-egg–laying gene. Sold commercially, these birds may lay eggs in shades of green, blue, brown, and pink and have any number of characteristics, but they are not recognized by any club or organization.

There's good news: This veritable "mutt" chicken makes a great pet. They won't breed true, nor are they appropriate for showing, but they are readily available from most hatcheries, are very friendly, and can surprise you with beautiful eggs.

Egg Lore

Did you know that blue eggs are ***not*** lower in cholesterol than brown or white? This myth, propagated by hatcheries to sell more chicks, has been hard to shake for decades. Sure, they're pretty on the outside, but blue eggs have the same nutrients and taste as brown or white ones, provided the birds were raised on the same feed. Diet, not shell color, dictates the nutritional value of eggs.

Chapter 4

The Family Chicken

Chickens and children are like peas and carrots—they just make sense together. Chicken-keeping chores are easy and routine, making them perfect for youngsters. There's an appropriate task for every age, from toddlers to teens. Kids and chicks thoroughly enjoy each other's company, and the presence of a small flock can really enhance a family's dynamic. Just ask Jonah Hauser (pictured opposite), a plucky nine-year-old who lives with his parents, two brothers (one of whom is off to college), and flock of eight beautiful hens. Jonah has helped tend his family's flock for more than half his life, and he knows each of his ladies' unique personalities, snack preferences, and laying habits. He collects eggs, feeds them spaghetti by hand, and occasionally snuggles with them (especially Mustard, a particularly cuddly and affectionate Welsumer hen). Jonah's mom, Danise, swears that the flock builds community and knits the neighborhood together. With the only flock on their block, the Hausers share their abundance of eggs with neighbors, left and right. Young children come to visit the Hauser flock and pick choice eggs from favorite hens. Neighbors say they enjoy the melodic sounds of the hens' clucks and coos, and the residents of a nearby apartment complex say they feel as if they live deep in the country. The family has never had a single complaint—quite the contrary. In fact, they're sure they'd get complaints if they ever got rid of their flock.

With the resurgence in the popularity of the family flock, the Hausers are just one example of the thousands of families that are enjoying the quaint comfort of chickens. Of course, enjoying a family lifestyle with chickens should be fun and easy. The addition of chickens to a family should not cause undue stress or tension among family members. This chapter is designed to help parents, caretakers of children, and even teachers and other community members interested in establishing a communal flock find the joy in keeping chickens—together.

Raising Children, Raising Chickens

Many adults have a thing or two to learn about where our food comes from. Decades spent transitioning away from an agricultural life have left several generations of Americans ignorant about the very products that sustain their health. Family flocks can begin to turn that around.

Learning Where Food Comes From

Children who are raised around chickens learn *not* to take their food for granted. They see firsthand how hard their hens work to lay a single egg, and if they are involved in creating budgets and spending money on feed, they will come to recognize just how much money is involved in producing one healthy egg. For these children, eggs are not faceless, white orbs to be thrown into a shopping cart; they are the result of hard work from beloved family pets that have been cared for with intention and mindfulness.

Learning How to Care for Others

One of the best lessons to be learned from a flock of pet chickens is the responsibility of caring for another being. While chickens are relatively easy to care for as far as pets or livestock go, they have certain needs. Putting their health and well-being in the hands of an older child or teen can foster a sense of accountability while making it fun. Older children can be responsible for daily and weekly chores, bird handling, preventive care, and making vet appointments as necessary, for instance. This relationship between young adult and a flock of chickens may set the stage for responsible actions and choices later in life.

Learning Fiscal Responsibility

Chickens are the perfect pets to teach children that hard work pays off. For many animal-loving children, just the care and enjoyment of keeping a flock of chickens may be enough to perpetuate the relationship. However, the bonus of eggs is certainly a boon. Older kids and teens can use those eggs as a product to sell or trade, teaching valuable lessons about the worth of certain commodities. Moreover, when your child works with egg-buying customers, he or she is learning basic manners and customer service skills. Early lessons in fiscal responsibility are probably the greatest lesson of all. If your child takes on the selling of your family's surplus eggs, he or she will learn how to promote the product and gain valuable experience handling money, saving it, and spending it accordingly.

Jonah, age nine, holds Mustard, a beloved Welsumer hen. Jonah lives with his family and flock in Asheville, North Carolina.

Learning the ABCs

What do kids and chickens have in common? They are both pretty messy! Thankfully for parents, there are a few easy ways to keep those messes from combining and becoming overwhelming, indoors or out.

Following the Rules

When your children partake in chicken chores, or simply visit the coop to play with their favorite hen, consider implementing a few of these rules:

1. ALWAYS wash hands after handling chickens. This is the only rule that should be non-negotiable. Simple as that.

2. Designate a single pair of shoes or boots for each family member to be used exclusively and only for trips to the coop. Choose a material that is easy to hose off and choose a style that is easy to pull on and take off. Skip the laces. A favorite choice is a low pair of muck boots, rain boots, or Wellies. Let young children choose their boots and remind them that they are to use them only as "chicken boots."
3. Remove or change soiled clothing after handling birds. Chicken feet are arguably the dirtiest part of a hen, and there are a few ways to pick up a chicken that don't include intimate contact with those piggies. (See page 159 for one recommended chicken-handling method.) Large-breed, fluffy-bottomed hens often have droppings among their rear feathers as well. To keep the manure from spreading on clothing, young children may enjoy wearing an apron designated for chicken holding/hugging that is easy to remove without having to change clothes fully.

Designate a special pair of "coop boots" for each member of the family.

4. Use an egg basket to collect eggs. This rule will turn up again under "Peeps 'n' Pipsqueaks" (page 62), but it's worth mentioning here as well. It's important to keep eggs contained to baskets during transit from the coop to the kitchen simply because eggs are very fragile. A broken egg is a loss, but it's also a sticky, inconvenient mess to clean up.

Handling Chickens

Just as there must be rules around hygiene related to chickens, it's wise to have a few rules in place for the handling of your chickens—for both the chicken's safety and the children's (but mostly for the chicken's).

The unique needs of very young chicks will be covered in chapter 5, but for now, it's enough to know that these tiny birds are very delicate. They have petite frames, tiny bones, and only a thin layer of light down feathers covering their bodies. Squeezing or roughly grabbing a young chick could cause serious and crippling injury. For this reason, it's gravely important to teach children, especially very young children and toddlers, to handle the birds gently and only and *always* with adult supervision. Being animals of prey, chickens are instinctively a little put off by the idea of being held and tend to be rather wiggly when in arms. Because a squirmy chick is an unpredictable chick, a firm grip is needed to keep the chicks from wiggling out of your (or your child's) grasp and falling to the floor. It may seem like two opposing suggestions are being offered here, and they are: Chicks should be held firmly enough so they do not fall but gently enough so they are not squeezed and hurt.

Adult chickens, on the other hand, are far less fragile. They can easily tolerate being held by adults and children alike. However, the same rules apply: Some chickens just don't appreciate being held. For those wiggly, squirmy birds, a very firm grasp of the wings is imperative to getting a good hold. Otherwise, the handler runs the risk of getting a wing flapped right in the face (or worse, in the mouth or eye). (The proper etiquette for holding and handling a grown bird is covered in chapter 9.)

Chasing Chickens

It should probably go without saying that chasing chickens is *not* a recommended practice, for adults and children alike. A chicken's natural flight response is a sharpened tool designed to protect them from predators. When a child goes running toward a hen for a hug, the hen sees only a creature that (she thinks) wants to eat her, not an affectionate child who loves her. Chicken chasing is practically a childhood pastime, but if you want your hens to trust you and your children, save yourself and your chickens the stress by creating a "No Chasing Chickens" rule for all youngsters who care for or visit your coop.

Wrangling Roosters

When it comes to children and chickens, roosters are a mixed bag. Some roosters are incredibly docile and find children to be no threat at all. Those roosters are positively cuddly and love to be held and stroked. The vast majority of roosters and cockerels, however, are rather aggressive. Their aggression is natural and, in the chicken world, actually considered a good thing. For children, the risks that roosters pose often outweigh the benefits. Roosters are surprisingly strong for their size, have sharp spurs, and tend to attack when your guard is down and your back is turned (very literally). A single encounter with a mean rooster is enough to terrify and put off a young child from birds for life. Unless you have a great deal of space or a reliably meek rooster, keep your family flock to hens only, at least until the kids are teens and older.

Chickens raised respectfully around children tend to be curious, easygoing, and tolerant of human contact and handling.

The Neighborhood Chicken

Chickens make great additions to neighborhood communities as well as to individual backyards. Starting a neighborhood flock, rather than a privately owned, independent flock, has so many advantages.

Starting a Communal Flock

Going into chicken ownership with other families is a great way to build community and is well worth considering if you or your family has shied away from some of the responsibilities that keeping a personal flock

Tales from the Coop

When our daughter was a baby and just learning to pull herself up, she would crawl over to the window to pull up and peer out at the chickens. She would squeal with the purest delight every time the rooster crowed. We would read her a sweet peek-a-boo–style book called *Grow Up*, which featured several young animals and their adult counterparts, including the chick and the chicken. We made the animal noises for each one, and hands down, the sound of the chicken was her favorite (bok, bok, bok, ba-CAWK).

Then later, one rainy afternoon, she pulled up to the window, pointed at the chickens, and shouted her first animal sounds: bok! bok!

entails. More notably, sharing a backyard flock with other families divides the work, the costs, and the responsibilities of chicken ownership.

Cinsider the following four things if you're interested in starting a community flock:

1. First, gauge interest by talking to other families or individuals in your neighborhood and find out who might be interested in keeping a shared flock of chickens. Ideally, start small and limit the involvement to two or three families other than your own; fewer if the families are very large. The reason for this is to keep the flock's size manageable. The larger the number of people the flock will need to feed, the bigger the flock will need to be and the more space the coop will command. Also, try to collaborate with families that live within walking distance. This will make chores easier for everyone.
2. When you have several interested parties, set up a meeting to discuss particulars. Collectively, you'll want to identify each individual's strengths and preferences. Then, divvy up the responsibilities based on skill set and personal interest. For example, tech savvy teens may enjoy researching chicken breeds on the Internet, a handy individual may take on the task of designing the coop, while others may take on the job of contacting the city and getting all of the legal ducks in a row. Designate a responsible individual to handle money, and decide how much each family must contribute to support the start-up costs. Give everyone a task to keep jobs evenly distributed, or break up into small groups to get started.
3. At the next meeting, set some dates. Schedule a date to build the coop collectively—all participating families should be in attendance to help. As a group, you'll need to decide how many chickens will realistically supply each family with enough to feed them weekly (as a rule, one or two hens per person is usually enough). Come to a decision on breeds and set a date to purchase chicks or pullets. As a group, decide who is best equipped to care for young chicks and will brood them, and on whose property the coop will be built.
4. Within the group, create a schedule of chores and choose responsible parties. (Daily, weekly, monthly, and yearly chores are listed in more detail in chapter 11.) My suggestion is to rotate fun and easy chores, such as egg collecting and feeding among families, and then collaboratively tackle the bigger chores, such as the yearly coop cleaning.

To be successful, a community flock relies on equal, joint effort by multiple families or individuals. For a shared flock to work, everyone must do his or her fair share of work, pitch in, and be reliable. The excitement for joining together on this venture should be mutual. Only join forces with others who are enthusiastic, care about the chickens, and are willing to pull their weight.

Starting a School Coop

A school coop is a great way to reach hundreds of kids (or more) and offer the educational benefits of animal care and responsibility, teach a new generation about where their food comes from, and, of course, take the classroom outside for hands-on learning. A small chicken coop is the perfect complement to a school garden, too. Students will learn about recycling nutrients, soil health, composting, natural life and growth cycles, and so much more. Alone or in conjunction with a school garden, a chicken coop can support nutrition and wellness initiatives and environmental education in a fun, outdoor environment.

The same rules that apply to home flocks and communal neighborhood flocks apply to school flocks. Housing should be erected prior to purchasing chicks, and all other equipment and research should be thoroughly prepared by whoever is spearheading the project. As with communal flocks, it helps to have a small team dedicated to the success of the flock that meets regularly, delegates tasks, manages funds, and cares for the birds. There are a myriad of ways a school can tackle chicken keeping, and how your individual school does it will depend on factors such as location, the school's budget, and regulations based on whether the school is public, private, or a charter. You may even find local businesses that are happy to donate supplies, building materials, volunteer time, or even donate a few starter chicks to see the project through. Starting a school coop is the essence of community. With just a little help from dozens of sources (or more), a school flock can be supported without any one individual or organization fronting the majority of the work or costs.

School-based chicken flocks also offer fun activities for students, no matter their ages. Chores, tasks, or chicken-related projects can be divvied up among the various grades, with assignments based on age appropriateness. Here are a few ideas to get all grades involved:

- Incubate and hatch eggs.
- Raise chicks in one of the classrooms.
- Hold a student-based naming contest for all but one of the birds. Put the naming of the final chicken up for auction to benefit the school.
- Collect, wash, and package eggs.
- Decorate egg cartons through painting, collage, and other media as appropriate to the ages of the students.
- Sell eggs to offset coop or feed costs, teaching children to budget and work for goods.
- Work with older students to present chicken-keeping training classes to new volunteers and families.

A dedicated volunteer team will be significant to a school coop's success. Even if the school is open year-round, seasonal breaks will leave the school empty, shy of a few teachers and administrative staff, and leaving few (or no one) to care for the chickens. Some schools use a rotating schedule of volunteers (teachers, parents, and administrative staff) to provide daily care when school is out; other schools choose to relocate the flock to a family's backyard or local farm during breaks. Again, how your school approaches these tasks will be determined by the choices of your chicken committee, the availability of its members, and the location of your school. Even with a little extra legwork and coordination, many find that the educational benefits of a school-based coop far outweigh the extra effort.

Peeps 'n' Pipsqueaks: Chicken Chores for Young People

The chores and maintenance that chickens require are perfect for children and teens. The tasks are easy, straightforward, and very routine in nature, and the rewards (egg fun! chicken antics!) are like no other. But if you're looking for efficiency, you won't always find it when giving chicken chores to young children. Kids like to take their time. Physically, they are still refining gross motor skills and—this may sound obvious—are very small. The extra time it takes out of *your* day to teach and guide them, however, is far outweighed by the benefits your little one(s) experience from the work. Kids love to help out and to feel as if their work is important to you and the greater family or community. When children believe their efforts matter and that they can make a difference, it can boost their self-esteem. The responsibility of caring for chickens can instill confidence that will linger for a lifetime.

Below are a few easy children-chicken chores, in no particular order.

Collecting Eggs. This is possibly the most rewarding chicken chore of all, and you may be hard-pressed to give it up to your kids. Escort very young children to the coop and back and help them along the way. Provide low nest boxes for your kids and step stools if they still have trouble reaching the boxes. Make the experience fun and comfortable for them. As mentioned earlier in this chapter, encourage the use of a designated egg basket for each collecting trip to protect the eggs and organize the process.

Giving Scratch. When the days turn cooler and autumn rolls around, a little extra scratch will help your hens bulk up for the coming winter. Don't be fooled into thinking this is a throwaway chore. Children love the thrill of calling to their pet hens and enjoy watching them waddle over and gobble up scratch—even out of their hands. The way to a hen's heart is through her stomach: Giving scratch and other treats helps build trust between the flock and the kids.

Refilling Feeders and Supplement Troughs. If your young child has mastered the ability to pour and handle small bags or buckets of feed and supplements, let them feed the flock and replenish supplements. Young children will enjoy the hens' excitement but will still require adult supervision. This is the perfect daily chore for older children and teens who are learning about responsibility and care for others.

Selling Eggs. Arguably one of the more fun chores, selling eggs may include a marketing campaign, building negotiation skills, and handling the exchange of money. Support your child if he or she chooses to build a neighborhood "egg stand" and sell eggs by the half or whole dozen. Help them build the stand, design and paint signs, and then spread the word through your community before the stand goes up.

Children feel empowered when they are given chores appropriate for their age and development.

Learning to care for animals from an early age builds a sense of respect and responsibility for the natural world.

Of course, the older your child is, the more complex chores they will be capable of handling. Older kids and teens can lend a hand when it comes to coop cleaning, checking individual birds for general health, administering medications when needed, and much more. They can conduct daily visual checks of the coop and enclosure, help with repairs, and keep an eye out for pests and predators. Give your child tasks based on his or her age and skill level. To keep them (and you) from getting frustrated, it's important to keep chores appropriate and achievable. At the end of the day, which chores you give your child is less important than the fact that they are participating as part of the family.

At age four, Mavis helps her moms, Cindy and Laura, by collecting eggs, occasionally herding chickens back into their enclosures and giving the wayward hen a hug or two. "We give Mavis more and more chores as she's developmentally ready," says Cindy.

Chapter 5

Starting Your Flock

So you've done the research and determined that backyard flocks are legal in your town. You've chosen a few breeds and sourced a reliable vendor for quality feed and supplements. The equipment is purchased and standing by. Your coop is built and at the ready. There's only one thing left to do: Add chickens.

Where will you begin? With fluffy, day-old chicks or point-of-lay pullets? Will you order your chicks from a hatchery or find a breeder? Or, will you take the road less traveled and give some old battery-cage hens a new home and some well-deserved freedom? There are many ways to acquire a small flock, and each has its advantages and drawbacks. Your choices may be affected by factors such as your location, how easy it is to find breeders or other chicken keepers, the breeds you have your heart set on, and your budget. This chapter will explore a few of the most popular ways to start a flock. As hatcheries are the easiest and most common way of purchasing chickens, we'll kick off the chapter with mail-order chicks in some depth. As the pastime for keeping chickens picks up steam, it's becoming easier and easier to find the fowl that suits any family's needs.

Chicks

Of all of the ways to start a flock of laying hens, raising them from chicks is the easiest. When compared to other methods (such as hatching eggs or buying adults, all of which this chapter will explore thoroughly), raising a flock from chicks is less expensive and more reliable, has more guarantees and more breeds to choose from, and chicks are way *(way)* cuter. No wonder the vast majority of chicken keepers choose to start their flock from chicks. What's more, the most popular way to get chicks is to order them through the mail from a commercial hatchery. Here's how it works.

The Magic of Mail Order

When flying on a plane, your journey has probably involved a combination of drowsy sleep and restlessness. Maybe you accidentally fell asleep on a fellow passenger, or another talked your ears off. Despite being crammed in this high-flying vessel for hours, you usually reach your final destination a little dehydrated, eager for a good meal and a warm bed. Otherwise, you're probably none the worse for wear.

If you purchase day-old chicks from a hatchery, they undergo a similar travel experience before reaching your door. Luckily, they have you to greet them, ready to prepare their warm brooder and offer fresh water and healthy feed to help them settle in. This method of selling and shipping day-old chicks is rather old: The U.S. Postal Service has been shipping chicks through the mail for more than 100 years and shows no signs of slowing. In that time, the marriage of the post office and poultry hatcheries has nearly perfected the art of shipping chicks. The vast majority of birds ordered in this way arrive alive and healthy. This is possible for several reasons.

Before hatching, an unborn chick absorbs the yolk of her egg, giving her the strength to hatch out of her egg and buying a few days' time to remain under the mother without additional food. In nature, this three-day window allows time for the remaining eggs in the clutch to hatch. In the hatchery, this window allows time for vaccination, "packaging," and expedited shipment to you.

Knowing that this window is critical, many hatcheries plan shipments on Mondays and Wednesdays to reduce the chances of the chicks sitting at the post office through a weekend. They calculate projections of which birds will hatch each week and based on those projections, the weather, and a customer's particular order, hatcheries will aim to ship an order of chicks as close to your chosen delivery date as possible. To safeguard the sensitive chicks against the cold, hatcheries have several techniques they employ, such as adding "packing peanut" chicks (additional birds, usually males, to supply added warmth) or heat packs to the order.

Opinions abound about the chick-shipping process. Some long-time chicken keepers swear by it; others prefer not to support such large hatching operations for a variety of reasons (most of which this book won't cover, but a quick online search will yield hundreds of

The U.S. Postal Service has been working with hatcheries for nearly 100 years to safely ship chicks to chicken keepers across the country.

results). For some new chicken keepers, purchasing hatchery chicks is the only way they're able to get started. My recommendation is always to fully educate yourself on the topic before making a decision.

90 Days to Delivery: Researching and Ordering

Before deciding if mail-order chicks are for you, see if the necessary planning will suit your schedule. When planning a flock from mail-order day-old chicks, work backward. Several months before your preferred arrival date, pick up a catalog or browse the Internet for your hatchery of choice. There are many available: Some specialize in rare or exotic breeds, while others promote heritage birds. Some hatcheries specialize in shipping orders as small as three to five chicks. You may choose based on any of these factors, or, from the hatchery's location, reputation, or simply their availability of your chosen breed. Very rare or popular breeds will sell out quickly, so it's best to order early.

Most hatcheries require a minimum shipment of 25 chicks to ensure warmth and safety.

Once you have decided on breed, consider if you would like your chicks "sexed." *Sexing, vent sexing,* or simply *venting* is a service the hatchery offers on the chicks you purchase to determine if the bird is male or female. Trained specialists called *chick sexers* learn to look at the vent of a day-old chick for a small bump that may indicate gender. With about 90 percent accuracy, the odds are actually pretty good. However, with that said, sexing chicks is more of an art than a science. To be prudent, have a backup plan should you accidentally get a rooster.

The vast majority of chicken keepers prefer females to males, so purchasing sexed pullets will come with a slightly higher cost. If you prefer not to pay the difference (or you're up for a bit of a surprise), hatcheries offer chicks *straight run*. Straight run birds are unsexed, but you will undoubtedly receive a mix of male and female birds—the ratio typically runs around 50/50. Even so, these estimates are not guaranteed, so ordering chicks straight run is a little like playing the chicken lottery.

Next, you'll want to decide whether to have your chicks vaccinated, or not, and if so, for which diseases. Many hatcheries offer vaccinations for Marek's disease, Newcastle, and several strains of coccidiosis. Decide before ordering, since most vaccines must be administered in the first day of life. Though vaccinations per chick may cost only pennies, do your research to decide if vaccinating is a route you want to take. (See page 134 for more on vaccinations.)

Last but not least, it's time to order. Once you've chosen a hatchery and selected your birds' breed, gender, quantity, and vaccinations, it's time to choose your delivery date. Hatcheries provide a delivery window of about a week. This window is partly determined by the availability of your breed in your chosen gender—females, of any breed, tend to sell out faster than males. Hatcheries do their best to estimate how many chicks will be available during any given week, but nothing can be guaranteed. Sometimes, eggs don't hatch out as planned, or they underestimate a breed's popularity. However, they'll be sure to send updates as it gets closer to hatch date. Remember, the sooner you place your order, the greater the likelihood you'll receive the chicks you want, when you want them.

When your chicks arrive, handle them gently and point them toward food and water in the brooder.

How Does Shipping Work?

Hatcheries have nearly perfected the art of delivering healthy peeping chicks to chicken enthusiasts around the country. The U.S. Postal Service will determine how long it will take for your order to arrive; from there, a hatchery will determine a safe minimum number of chicks. The longer the travel time to your location, the higher your minimum of birds will be. Smaller shipments during the colder months may require purchase of a heat pack for added warmth. Large quantities of chicks shipped in the heat, on the other hand, are sent in multiple boxes to reduce body warmth. A hatchery's shipping department will examine the weather at their location and at the chick's destination, and then pack them accordingly. Each individual hatchery may have a slightly different policy on shipping, so contact your chosen hatchery for details about shipping your chicks. (Hatcheries are listed in "Resources" on page 214.)

The Time of the Season

Ever wonder why "spring chickens" are the most popular kind? The mild temperatures of spring make it the most ideal season for hatching and shipping chicks across the country. Ordering chicks early in the year allows backyard chicken keepers to get eggs from their pullets before the year is out, too. Since it takes about five months for a pullet to begin laying, many like to order their chicks in January or February for a spring delivery and late-summer eggs.

Though many hatcheries will ship year-round, autumn is another favorable time of year to receive chicks through the mail. Like spring, the fall can be relatively mild in most regions, reducing the chance of losing birds to either extreme heat or extreme cold. Of the two, cold temperatures are a chick's biggest threat.

Traditionally, hatcheries have required a minimum of 25 birds in order to make a shipment. This magic number allows the day-olds to huddle together for warmth and

Countdown to Cuteness

Once you've chosen your breed of choice, it's time to determine the best option for starting a flock. Before ordering your chicks, make sure you have a plan and know what you want.

8 to 12 Weeks to Delivery Day: Choosing

Throughout the winter, keep tabs on your favorite hatcheries. Check out their websites or subscribe to their e-mail list for updates on when chicks are becoming available. Place your order for chicks as soon as your chosen breeds are available during the week you've selected for delivery. Rare breeds sell out rather quickly, and females of any breed sell out faster than males.

6 to 8 Weeks to Delivery Day: Nesting

Those tiny balls of fluff that you're expecting won't stay tiny or fluffy for very long. Chicks can be ready to live without supplemental heat in just about a month, so they really seem to grow in the blink of an eye. A few months before your delivery date, have your coop ready—or at least in the works. Remember that prefab coops may have waiting lists. It will save loads of stress and scrambling (pun intended) on your part later.

3 Weeks to Delivery Day: Shopping

With a few weeks to go, now is the time to go on a shopping spree: Purchase equipment such as a chick water font, chick feeder, brooder, heat lamp, bulbs, and more. (See "Chick Checklist" on page 82 for everything you'll need and chapter 6 for how to use it.) Don't leave anything to the last minute. Like the coop, have the brooder constructed and standing by. Get to know your local feed and seed stores and confirm that a local establishment is a reliable source from which to purchase chicken feed. Have a bag of starter feed purchased and standing by—you'll need it right away.

1 Week to Delivery Day: Setting Up and Doing a Dry Run

Now is the time to set up your brooder. As you'll see, a basement, laundry room, garage, or mudroom with electricity is an ideal place for a brooder. Choose a location that has easy access for quick feed refills, water changes, and chick checkups. Line the brooder floor with pine shavings and place a thin layer of paper towels on top. The paper towels prevent chicks from eating the bedding for the first few days as they locate their feed. You can remove it when the chicks are a week or two old.

The hatchery will provide you with a window of time within a week that your chicks may arrive—but rarely will you have an exact day. Be ready to get an early morning phone call (mine usually come at about 5:00 A.M.) from the post office any day of your scheduled delivery week. It helps to give your local post office advanced notice of your expected live delivery. They'll usually make note of the special circumstances and call you right away.

Several days before the first estimated delivery date, run the heat lamp and place a thermometer at chick height in the brooder. Heat is one of the most crucial elements to successfully raising small chicks. (More about this on page 77.) In your brooder setup, take into account that the lamp will need to be raised higher or lower to provide the right temperature for the birds. (Heat lamps are pretty old school: There are only two settings, on or off, changed by unplugging the lamp, so it will need to be physically moved to adjust the temperature.) Always keep a back-up bulb in case your primary bulb burns out.

In this first test run, make sure the brooder is able to stay warm both day and night. It should also be free from heavy drafts but offer adequate ventilation, keeping an even temperature without getting stuffy. Check the thermometer at different times of day during your test run to confirm that all is working as planned.

Delivery Day

The big day has finally arrived. Most likely, you'll hear the phone ring in the wee hours of the morning, and it will be your local post office asking you to pick up the chirping box with your name on it. Your new charges will be hungry, thirsty, a little jet-lagged, and, of course, ready to see a friendly face.

Once you get to the post office, take a peek inside the box. Immediately report any mortalities to your local postmaster, and make a note to contact the hatchery as soon as you arrive home. Loud peeping and chirping is a good sign—it means the chicks are strong, healthy, and robust!

When you arrive home with your charges, fill their waterer and feeder immediately. Remember, the brooder should be warmed to 95°F (35°C) at chick height. As you remove the chicks from their shipping box, one by one, put each bird's beak in the fresh water of the font and place her down beside it. This process encourages hydration and helps each bird locate the water source. Within a few minutes, the chicks will be eating, drinking, and orienting themselves to their new home. Don't be alarmed if they pass out within the hour—they've had a very busy first few days of life and need the rest.

Once everyone is settled, pay close attention to the chicks' behavior. If they are huddled close together under the heat lamp, they are too cold. If they are mingling around the periphery of the brooder, as far from the heat as possible, they are too hot. A perfect brooder has chicks cheeping contentedly, walking and pecking around, moving among the food, water, and heat sources evenly.

Finally, resist the temptation to handle the chicks too much on the first day. Instead, allow them time to rest, and pick up the camera to take a few photos. In just a week, they will have already morphed into different creatures, so capture the cuteness of the early days while you can.

While the success rate of shipping chicks is very high, there are a few risks that are important to know about.

produce enough heat to survive the journey. These days, hatcheries understand that 25 birds are a lot for the backyard hobbyist. As such, more hatcheries are offering smaller shipment sizes, such as 10 to 15, and/or supplemental heat packs for small orders at an additional fee.

The Pros and Cons

Ordering chicks from a hatchery can be a safe and efficient way to receive chicks, but there is always a chance that you will receive birds that were injured in transit, failed to thrive, or died during shipping. The process of travel can stress already weakened chicks, and some hatchery chicks almost always have pasted vents, a potentially fatal chick ailment. (Luckily, there's an unbelievably simple fix for this; see page 80 for the "pasty butt" remedy.) It's best to be prepared for these scenarios and understand that death is a possibility if you buy chicks this way.

As far as getting to choose your chicks, buying through a hatchery allows you to try a variety of breeds and brood them all at the same time. It's also quite convenient to pick the week you'd like to start your flock if you travel or have children. Unlike local breeders, most hatcheries will either replace the unwanted males you receive with females, or they will credit the purchase price. The same goes for chicks that do not survive travel or die within a day or two of arrival. Also, hatcheries offer vaccinations for day-old birds if that is a priority to you.

Shopping Close to Home

Fortunately for small breeders, hatcheries don't have the monopoly on day-old chicks. Of course, some breeders will ship their birds to customers across the country, but shipping policies vary widely. If you would prefer not to have your birds shipped or would rather have the experience of picking them out yourself, young chicks can also be purchased from local

breeders or farmers. Getting to choose the family flock by hand is especially meaningful for children, creating memories and establishing bonds with your birds from the first.

To locate breeders in your area, first ask around. Check at your local feed and seed store (if there is one) and get recommendations. See who comes highly praised for healthy birds and good management practices; don't underestimate good word-of-mouth recommendations. Another great place to look is online: Check chicken-keeping forums for backyard breeders and the American Livestock Breeds Conservancy for registered poultry breeders in your region. Be prepared to travel a bit to find quality breeders, and don't expect that they will have a large variety of breeds when you get there. Poultry fanciers often focus on one or two breeds, preferring quality to quantity.

When inquiring about purchasing chicks from breeders, don't be afraid to ask questions. In fact, there are a few things you absolutely must learn from breeders *before* buying their birds, including...

- Whether the chicks are sexed or straight run. Many breeders are unable to sex chicks, and by purchasing their birds, you may (and very likely will) end up with a few males. The same rules apply for hatching rates with breeders as with hatcheries—expect roughly a 50/50 ratio of males to females with all straight run chicks.
- Whether the chicks have been vaccinated. Most vaccines must be administered in the first day of life to be effective, so it is up to the breeder to administer the inoculations upon hatching.
- What management practices they follow. Ask to visit the breeder before buying your birds. As with purchasing adult chickens (more on that to come), only buy birds from breeders or farms with good biosecurity measures in place. The coop shouldn't smell, and birds should appear healthy, with bright eyes, combs, and wattles. Wear shoe covers or change shoes between visits to different poultry establishments. Use your eyes, ears, and good judgment and bring home only healthy chickens.

The Pros and Cons

Buying your chicks locally reduces your carbon footprint and helps to support small local farms, breeders, and businesses. Chicks purchased locally endure far less travel stress and are less prone to pasted vents (although it doesn't entirely eliminate the chances of pasting up). On the other hand, locally purchased chicks are rarely vaccinated and rarely sexed.

Farm Supply Store Chicks

Another source for day-old chicks in the spring is the large chain farm supply stores, pet stores, feed and seed stores, and garden centers. If you choose to purchase chicks from these vendors, ask about the birds' origins before buying. The majority of chicks sold through these vendors are purchased in large quantities from hatcheries and shipped to the stores (and thus, aren't locally bred birds). Ask to see the paperwork that shipped with the chicks to establish their origin, confirm whether they are sexed or straight run, and determine whether they have been vaccinated. If you choose to purchase the chicks, ask for copies of the shipping information for your records.

Watching a newborn chicken hatch in an incubator is an exciting and educational part of raising chickens.

The Pros and Cons

The largest drawback to purchasing chicks from big box farm supply stores is that their origins are largely a mystery. This is why it is so important to see any paperwork that may be available. While these chicks are very easy to come by and usually rather inexpensive, farm supply store chicks may not be in best health; they may have been handled repeatedly by customers (read: excited children) and potentially stressed from the store's environment. These birds are often brooded in large tubs or containers with only a few waterers and feeders, so it can be unclear whether all chicks are able to find their food and water source. If you choose to buy these chicks, follow the recommended guidelines on picking out only the healthiest birds.

Fertile Hatching Eggs

Hatching fertilized eggs in a home incubator is an incredibly educational and exciting experience, for children and adults alike. There's nothing like the thrill of anticipation, watching and caring for the eggs daily. Come hatch day (or days), though, you may experience sheer bliss as your new charges make their way into the world, or there may be heartbreak as some birds fail to thrive before your eyes. Some eggs simply won't hatch at all. Rather than feel discouraged about the reality of home hatching, it's best to carefully examine the many variables that go into hatching fertile eggs. If shipped, will they arrive intact? How many will successfully hatch? Of those that hatch, how many will be male or female? Preparing yourself for the many potential surprises that hatching can bring will help you find success through this method of acquiring chickens.

First, fertile hatching eggs are about as easy to come by as chicks. Local farms or breeders may be willing to sell fertile eggs from some of their birds. Be sure to ask if the breed of rooster that fertilized the egg was the same breed as the hen that laid it; otherwise, you may end up with "mystery" chickens of mixed breeds (which can be fun, too). Also, some of the very same hatcheries that will ship day-old chicks are able and willing to ship fertilized eggs for hatching. Like chicks, shipping fertile eggs requires extreme care in transit, consistent temperatures, and your availability at pick-up time. Even if the shell appears intact and unbroken, the internal makeup of a chicken egg is rather delicate, and there's always the risk of damage during shipment. In addition to these obstacles, hatching a healthy chick from a fertile egg requires precise temperatures, just the right humidity, and sanitary conditions in the incubator. Being off just a little can largely impact the egg's viability. For these reasons, hatcheries simply can't guarantee hatch rates of their fertile eggs.

Chicken Lore

Classic folklore tells us that roosters hatch from pointed eggs and hens hatch from rounded eggs. It's doubtful that there is any truth to this myth; however, it's important to take the underlying message to heart for those hatching eggs: Incubate only normally shaped, clean, uncracked eggs. Sadly, oddly shaped or pointed eggs will not encourage proper embryo or chick development. Perhaps this myth was created to encourage chicken keepers to hatch only the most viable eggs.

Another variable that potential hatchers must face is that hatch rates can seem low to new chicken keepers: The hatch rate of shipped eggs is about 50 percent. This rate increases if you purchase the eggs from a local farm or breeder. Better yet would be fertilized eggs left under a broody hen to hatch naturally (and, thus, never moved). Before buying the eggs, it's important to thoroughly research how to incubate your own eggs and prepare yourself and your family for the work that goes into it.

The next variable is gender. Of the eggs that do hatch, expect some males. Despite the old wives' tales about egg shape or time of year, there is no guaranteed way to determine the gender of a chick while still inside the egg. Keep this in mind if you aren't able to keep or easily rehome any roosters.

The last consideration is cost. Monetarily, the cost for fertile hatching eggs is usually much less than purchasing adult birds but quite comparable to the cost of day-old chicks.

The Pros and Cons

Hatching your own chicks in an incubator is all about mystery: You don't know how many eggs will hatch, and of those that hatch, you don't know how many will be male or female. When you compare how many shipped eggs successfully arrive *and* hatch at home to the low mortality rate of shipped chicks, the true costs seem in favor of starting a first flock from live birds. Simply put: Shipping eggs is a risky venture, and hatching eggs offers very few guarantees. Most new chicken keepers prefer to get started with healthy, live chicks the first time around and leave hatching adventures for further down the road.

Feeding and Care

However you come to find young chicks in your charge, your duties as "mother hen" remain the same: You'll need to provide them with warmth, safety, shelter, food, water, and diligent care. Though their adult counterparts are rather hardy, baby chicks need a bit of coddling during their first few weeks of life. They're small, rather delicate creatures with very specific needs; they'll rely on you to keep them safe and offer them everything they need to grow and be healthy.

Brooding

A good brooder doesn't need to cost a pretty penny. Just about any secure vessel can double as a chick brooder: A large plastic storage bin, an old clawfoot tub, a kiddie pool, or even a sturdy cardboard box will do the trick. The chicks won't mind what the brooder is made from as long as it keeps them warm, safe, fits their feeder and waterer, and holds them all comfortably. From the get-go, make your brooder large enough for them to grow into but not so big that areas get very cold. A good rule of thumb is to give your chicks about 2 sq. ft. (0.2 sq. m) of space per chick. It will seem like a lot of room at first, but the little fluff balls will grow into it.

Line the brooder with several inches of pine shavings. For the first few days in the new brooder, layer paper towels over the bedding to keep the chicks from pecking or eating the shavings. Once they have established the location of their food and water, you can remove the paper towels. To ensure that the containers, and their contents, stay cool, do not place feeders and water fonts directly under the heat lamp. This is especially important for metal containers.

Another consideration for brooding chicks is location. Most notably, chicks (like their adult counterparts) scratch at their bedding and feed, kicking up a fine dust that will

Flying the Coop

Unlike adult chickens, young chicks are pretty adept flyers. Don't be surprised if you walk up to the brooder one day and find a gaggle of gals perched on the brooder walls. For their safety (and your sanity) the easiest remedy for new flyers is to place a layer of bird netting or chicken wire over the open brooder. As a peace offering to your new pals, give them small sticks to perch on instead; thin tree branches, hoisted about 5 to 6 in. (13 to 15 cm) from the ground, work beautifully. The chicks will enjoy honing their roosting skills, and you'll save yourself a lot of time and mess.

settle on every surface of the surrounding area. This is a significant consideration when finding a location to brood them. At all costs, keep the brooder out of places where food is prepared or stored. Garages, mudrooms, or basements all make suitable shelter as long as temperature is accounted for. Sheds, barns, or large chicken coops will also work beautifully as long as they meet the chick's needs of safety and warmth, are free from drafts, and it's easy for you to check on them several times a day. If you must brood your charges inside your home, choose a room that is easy to clean top to bottom, such as a spare bathroom or laundry room.

These newly hatched chicks are staying nice and warm. For the first week of their lives, chicks should be kept in a brooding area that is 95°F (35°C).

Finally, have your brooder set up and warmed before your chicks arrive. "Countdown to Cuteness" on page 70 provides you with a week-by-week checklist and walks you through a dry run.

Heat

Achieving the correct brooder temperature is likely the most critical element to successfully rearing chicks. In fact, not providing adequate heat could mean the death of young chicks. For the first week of life, chicks require their surroundings to be 95°F (35°C); during the second week, they'll need temperatures of 90°F (32°C); during the third week, 85°F (29°); and so on. In other words, the ambient temperature should decrease by approximately 5 degrees per week until you reach 70°F (21°C), or, until the chicks are fully feathered. Adolescent poultry can be moved to outdoor housing at this time as long as the nights remain above 55°F (13°C) consistently in your area. (This is another reason to time your purchase of chicks for mild weather, such as the spring.) Most chicks are fully feathered and ready for outdoor living at five to six weeks of age.

The best method to heat a brooder is by using a poultry heat lamp with a 250-watt bulb. Even the new models sold today are rather simplistic. The heat lamp turns on by plugging in and has only one setting: on. To turn it off, simply unplug it. Coupled with an infrared 250-watt heat bulb and suspended above the brooder

(either hung or clipped), a single brooder lamp will warm small clutches of chicks nicely. Fortunately, if you need two, they're very reasonably priced (about $12 per lamp). When shopping for a heat lamp, you may notice styles with and without a guard on the brooder: *Always* purchase one with a guard. If the lamp should fall into the brooder, the guard will protect the bulb from breaking and prevent the lamp from coming into contact with the bedding. Improperly installed heat lamps most certainly pose a fire hazard, so always follow the manufacturer's recommendations for hanging or clipping. Some poultry supply catalogs also sell lamp stands for just this purpose.

When it comes to brooder temperature, it's best not to rely on "feel" alone. A poultry-safe thermometer placed at chick height will monitor the temperature and receive an accurate read. The temperature inside the brooder is often affected by a variety of factors, such as its location (is it under a sunny window?), the opening and closing of doors (is it in a heated or unheated room?), and the time of year. A brooder may heat up rather quickly in a small space, for instance, even with the lamp on the other side of the room. In a cold, drafty room, on the other hand, a few chicks may need two lamps. Using a thermometer eliminates guesswork and will help you troubleshoot heat issues quickly.

Newly hatched chicks need access to plenty of fresh water and starter feed. Your kids will love helping to keep chicks' fonts and feeders well stocked.

Finally, let the chicks be your guide. If they are huddled together under the heat lamp, they are probably cold. If they scatter to the periphery of the brooder, getting as far from the lamp as possible, they are probably too hot. Comfortable chicks will move evenly around the brooder, contentedly peeping, and eating and drinking at their leisure. Even if the thermometer reads the correct temperature for their age, it's best to follow the chick's lead and raise or lower the lamp accordingly.

Food

Baby chicks are eating machines. Their bodies are designed to grow incredibly fast, so they spend much of their time eating and sleeping. Within their first three weeks of life, chicks nearly double in size. In addition to growth, young chicks are hard at work replacing their fluffy down with real feathers, which requires a substantial amount of protein and takes a large amount of energy.

Chicks must have access to as much feed as they want 'round the clock. The best practice is to simply keep chick feeders full at all times and refill as you check on the chicks throughout the day. You may find that the little eating machines devour an entire feeder's worth of food in a day or even in just a few hours, depending on the number of birds in the brooder, of course. This is okay. Chicks will never overeat; they're growing after all. The feed you choose for baby chicks should be designated for their young age. This feed is often labeled Starter or Grower, and there are also medicated starter feeds to prevent against coccidiosis.

Tales from the Coop

If you have them readily available, or can easily source them, strongly consider dropping a few marbles into the lip of your day-old chicks' water font. On the morning our first chicks arrived, I sat at the edge of the brooder watching a tiny Black Australorp chick (who would later grow up to be Eleanor Rigby), drinking water. As all babies do, she became progressively sleepier the more she drank, eventually passing out with her entire beak submerged in the water. I waited a beat, thinking she would pop up when she couldn't breathe, but she didn't budge. I moved quickly and plucked her from the water. She was startled and unhappy to be awakened, but, thankfully, she was very much alive.

In the place of marbles, clean rocks or pieces of gravel comparable in size will work well to provide a shelf in the lip of the water font. This barrier will help to prevent chicks from drowning until the little ones grow out of their newbornlike sleepy phase.

Water

The same rules apply for fresh water as for feed. Refill the water font as needed so that the chicks have access to fresh, unsoiled water at all times. You may find that the first place chicks learn to roost is on the top of the font, often soiling the water below. As with adult chickens, it helps to raise the feeder and water font to sit at the back height of the chicks. With rapidly growing chicks, you'll likely need to reposition the font a bit higher each week. I've found that landscaping bricks work perfectly for this purpose, since they're very sturdy and not apt to tip. Finally, place a few marbles or clean rocks in the lip of the waterer to prevent drowsy chicks from falling asleep and drowning.

Safety

Chick safety falls into two categories: internal risks and external dangers.

Internal Risks. There are several areas of concern that pose internal risks to small chicks, including illnesses, such as coccidiosis, and other physical ailments, such as pasted vents. The parasite *Coccidia*, which cause the illness coccidiosis, may be treated preventively with either vaccination or medicated feed. Do your research and choose which, if any, you will administer to your chicks. If you decide to medicate, pick either the vaccine *or* the medicated feed, but not both: Used together, they will render each other ineffective.

Another common concern for chicks is pasty butt. Also called pasted vents, this condition occurs when feces dry and cake over the vent, preventing the bird from passing more droppings. Pasty butt is rather common with day-old chicks in their first week of life, especially if they have been shipped through the mail. The name warrants a giggle, but unfortunately, pasty butt is no laughing matter. Otherwise healthy chicks with unresolved pasted vents will certainly die. During the first week of life, check all chicks several times per day for pasted vents and use this simple remedy if you encounter an afflicted chick:

1. Warm and wet a few old washcloths, rags, or heavy-duty paper towels.
2. Move all chicks into temporary housing, such as a cardboard box (no need to heat it; they'll only be there as long as it takes to check all the birds).
3. One by one, pick up each chick and check her vent. If you see feces pasting the vent and blocking the hole, gently remove it with the washcloth. Take care not to pull on the downy feathers around the vent; a chick's skin is very thin and delicate.
4. Place each checked and/or washed chick back into the brooder. Repeat the process until you've checked all of the birds and they are all back in the brooder.

The sources of both coccidiosis and pasted vents are covered in more detail in chapter 9.

External Dangers. Small chicks are at risk for a variety of external dangers that include the cold, a number of predators, overcrowding, environmental stressors, and excessive handling. Since brooder temperature has already been discussed, this section will focus on other external dangers.

Many animals pose a threat to young chicks just as they do to adult birds, with a few additional predators because of the chick's small size. Care should be exercised to protect your brood from these predators. If the brooder is located in a barn, chicken coop, shed, or other outdoor structure, predator-proof it the way you would a chicken coop. If the brooder is located inside a human dwelling, such as in a basement, laundry room, or garage, your predators will likely be more manageable but not entirely eliminated. Crafty predators, such as raccoons, can easily slip through pet doors or windows. Household pets, such as cats and dogs, also pose a significant risk to young chicks. Take care to keep them separate from your chicks unless dutifully supervised. As an additional precaution, consider placing a barrier of chicken wire above the brooder. This barrier will also come in handy when the chicks learn to fly, too. Finally, don't underestimate the gumption of rats or mice. Even if they don't currently occupy your home or coop, the smell of the brooder could attract them (as if you needed another reason to keep the brooder clean).

Another risk to young chicks is environmental. Overcrowding and other stressors can lead to unpleasant situations, such as trampled chicks, chronic pecking, or injured birds. Like a coop, the brooder should have adequate space for the birds to move about freely, reach food and water easily, and for submissive birds to seek refuge from alphas.

Cutting the Cord

If you see a thin, black string hanging from the rear of a new chick, leave it be. This is the chick's umbilical cord. Some birds arrive with it still intact but not to worry: It will fall off within a few days. In the meantime, watch the chick for signs of bullying (other chicks may be attracted to it and try to pick at it). Since attached cords are unrelated to pasted vents, keep checking the chick for signs of pasty butt in addition to checking the cord.

Outside Time

Just as grown hens benefit from time on pasture, so do chicks. When the fluffy babes reach two to three weeks in age, and the weather cooperates (it should be above 70°F/21°C outside), give them a few minutes outdoors to peck at the grass and explore the world. They'll enjoy the new experience and the bright sunshine. Other than being good for the soul, early exposure to the biodiversity of your yard's soil will give your birds a healthy boost for when it comes time to transition outdoors permanently. Monitor the excursion and bring them inside before they seem chilly. Loud, alarmed peeping and huddling together are both signs they've become too cold.

As early as the first week, young chicks are already establishing a pecking order among their flock mates. Environmental stressors also include wire-bottom flooring in the brooder, which could lead to injury and/or bumblefoot; for this reason, follow the guidelines for bedding provided earlier in this chapter. Reduce excessive stimulation, such as that which occurs from using white lightbulbs. The constant harsh light can cause significant stress, won't give chicks a chance to rest, and will quickly lead to cannibalism. Instead, use the red lightbulbs designed specifically for poultry, often sold with other brooding supplies.

Finally, limit the handling of your chicks to just a few times per day. Moderate handling will build trust and get the chicks used to being handled by humans. Over-handling of very young chicks, whose small frames are rather fragile, could cause undue stress or lead to irreparable injury. They're cute, soft, and cuddly, and their contented cheeping is very sweet to listen to. There's no denying it: Chicks can be downright irresistible, especially to young children. However, handling should be exercised in moderation and always monitored by an adult. Create boundaries for visitors and children and stick to them. Particularly with kids, it's best to set firm ground rules for the handling of chicks. For instance, choose a time of day and a short time frame (say 10 to 15 minutes) for chick holding time (watching and talking to the chicks can go on for much longer). This gives the kiddo(s) something to look forward to, an opportunity to practice patience, gentleness, and respect, all without overwhelming the birds. And the chicks get a little sweet lovin', too. Everybody wins.

Chick Checklist

- ✓ Brooder (homemade or purchased).
- ✓ Heat lamp, with guard.
- ✓ Two red lightbulbs.
- ✓ Chick-sized water font, plus marbles (to prevent drowning).
- ✓ Chick-sized feeder.
- ✓ Starter feed.
- ✓ Chick-sized grit (if feeding fruits and vegetables).
- ✓ Pine shavings (for bedding).
- ✓ Paper towels.
- ✓ Washcloth and warm water (for potential vent pasting).
- ✓ Camera/phone (to capture the cuteness while you can!).

In the Beginning

Baby chicks are pretty needy: Think newborn baby, just without all the nursing and cuddling (well, there is *some* cuddling). Plan to clear your schedule for the first four weeks of chick rearing. As you already know, day-olds need constant food and water refills, bedding changes, vent cleaning (sometimes), daily checks (about five to six, roughly), and around-the-clock heat that changes slightly in temperature week-to-week. To give them the care they'll need in the beginning, postpone vacations or weekend getaways until they're fully feathered and in the coop. And if you're gone for long stretches of time during the day during that first month, enlist friends, family, or neighbors to do chick checks or schedule visits during lunch breaks to see how they're doing. Chickhood doesn't last forever, but it's a very important time in a chicken's life.

Adults

As you can imagine, raising chicks is not for everyone. It's time-consuming, and the chicks call for a lot of attention in the early weeks. They have special needs that adult birds don't, and raising layers from chickhood requires a good bit of patience before you'll see any eggs: about five to six months, in fact.

Pullet Together

Starting a flock from pullets, or adult female chickens, can be a great alternative for chicken keepers who are ready to dive right in and get eggs from the start. Some hatcheries sell "point of lay" pullets or young females who are just beginning to lay.

Adult chickens are more likely carriers of disease and illness, however, so if you embark on purchasing adults, you'll want to keep a sharp eye on health. To reduce the chance of infection and limit the spread of diseases or pests, bring home only ladies that are the picture of perfect health. Any hens you purchase should be alert, active, have glossy plumage, and bright eyes. (For more on what a healthy chicken should look like, see the description on page 130.) If you visit a breeder and their housing conditions are dirty, dim, dank, or smelly, buy your birds elsewhere.

The Pros and Cons

The greatest benefits to purchasing pullets are the guarantees they offer. With your birds fully feathered and grown, you'll know exactly what your birds will look like and exactly what breed they are. Best of all, there are no mystery males. Purchasing pullets will fast forward your chicken-keeping endeavors straight to the reward: the eggs.

On the other hand, starting with pullets does leave you with some uncertainty. Purchasing pullets requires a fair bit of research at the front, some travel to visit breeders or farms, and lots of questions. Ultimately, whether the birds were hatched on location or raised from hatchery chicks, you'll need to trust the seller that their birds are from the source they claim. How they were reared may also influence their temperament as adults: whether they are flighty and difficult to handle or docile and familiar with humans. With some pullets, you may not learn this until you bring them home.

Other Ways of Obtaining Chickens

Purchasing chickens from hatcheries, breeders, or fanciers isn't the only way to start your flock. There are many adult birds, or even whole flocks, who need a new coop to call home. Where do these birds come from? Well, not everyone who embarks on keeping chickens sticks with it for life (or even the lifespan of their birds). Some may start a flock only to find that keeping chickens just isn't for them. Others may jump the gun and start a flock before it's legal in their town. However it happens, there are occasionally flocks of healthy laying hens that find themselves homeless (or coopless, as the case may be). The good news for these gals is that they still have lives worth living and provide much in return. Mature hens often continue laying well into their twilight years, will continue to forage for pests and keep bug populations down, and will provide ample manure for compost until their last day. Giving these girls a second chance in your backyard could be a win-win situation.

Older Farm Birds. Traditional farmers will usually keep laying hens for much longer than factory farms do. While it's true that egg laying declines as a hen ages, many layer breeds will reliably give you eggs through their fifth or sixth year and beyond. Call ahead and ask farms if they sell their "retired" hens—get as much information as you can about the birds from the farmer or caretaker, asking such questions as: How old are the hens? What is their breed(s)? How are they managed (confined to coops, pastured, or rotated)?

Brooding Different Breeds

One of the best times to integrate birds of different breeds and sizes into the same flock is during chickhood. Chicks work to establish a hierarchy from the very beginning, and neither chicks nor adults discriminate on looks. Rather, they establish a pecking order on personality and assertiveness. Some breeds are known for having a docile temperament, and some are more aggressive. Of course, this influences an individual's personality. But, each bird is unique. Even if you had a flock of ten hens of the same breed, one bird would be alpha and others would be varying degrees of submissive. In other words, there's no need to shy away from mixing birds of different breeds because you're worried about kerfuffles. An individual bird's personality is more influential to the pecking order than her breed. You won't have to worry about managing the flock's hierarchy—they have it covered.

Poultry Swaps. Poultry swaps can be as large as county fairs or small events organized by regional poultry enthusiasts in a parking lot. If you choose to shop for your first birds at a poultry swap, come prepared. First, bring cash. Many folks looking to buy, sell, or trade won't accept credit cards, and some won't accept checks. Second, bring transportation for your birds. A large pet carrier or several small carriers works best; in a pinch, a cardboard box with holes punched through it will do. Third, arrive early and be ready to pounce when you see birds you want. Hens tend to go quickly, and there are almost always an abundance of roosters. Finally, thoroughly check any birds you intend to purchase for good health. Are their eyes clear? Do they have bright combs and wattles? How do the scales of their legs look? Check for signs of external parasites, injuries, loss of feathers, and other indicators of ill health. If need be, bring a checklist of what to look for (the description of a healthy chicken begins on page 130).

"Spent" Battery-Cage Hens. While morally fulfilling, starting a flock out of retired battery-cage hens can be heart wrenching and will probably pose health-related challenges that might be a bit much for first-time chicken owners. If you can find battery-cage hens for sale to the general public and choose to take them on, try to get as much information about them as you can from the organization. What were they fed? Were they given antibiotics or growth hormones? How old are they? Were they vaccinated? Stuff like that. Next, be prepared to offer these gals a little extra care and attention. They've likely spent their whole lives in a small cage, don't know how to forage (or even walk properly, for that matter), and have poor social skills (with humans and fellow chickens alike). Don't be shocked if they're missing feathers and are debeaked. If you choose to start a flock with some or all ex-factory-farmed hens, make sure you have established a relationship with a very reliable avian veterinarian who is willing to take calls. Finally, if your flock is to be a family flock, cared for and loved by children, you're probably better off getting birds another way. Ex-battery-cage hens will be skittish and scared and are usually breeds raised for high egg production, not temperament.

Shelter Rescues. The chickens that end up in county shelters are often found wandering by animal control. They may have escaped their enclosures in other backyard flocks or been set "free" by their previous owners. Sometimes, they are former battery-cage hens that have escaped the "farm." Some shelter birds are owner-surrendered, and if you're very lucky, these chickens will have been given to the shelter with a bit of background information. By and large, however, the origins of shelter birds will likely be unknown. Shelter staff may make educated guesses based on where the bird was found and if there are any farms nearby, but remember these are still guesses. If you feel comfortable with the mysterious pasts of shelter birds and choose to adopt one of these wayward gals, practice strict biosecurity measures. (See page 132 before integrating them with other birds.)

"Spent" Show Birds. Most award-winning show birds end up in the owner's breeding program or retired and doted on for their twilight years, so it's rather unlikely that breeders will want to part with these blue-ribbon gals. With that said, you may meet breeders who are looking to cull (remove) some healthy birds from their flock or breeding program. These may be excess birds the breeder is willing to part with or individuals that

Before the Chicks Come Home to Roost

There are several things you can do before your chicks arrive to smooth the way for a successful introduction into their new environment. First, establish a relationship with an area vet. Second, have a coop in place and ready to go. Finally, reach out to other chicken keepers in the area for advice and support.

1. **Establish an Avian Vet.** Chickens are hardy, scrappy creatures, and the truth is that they rarely get sick, only occasionally become injured, and when kept healthy through preventive measures live long, content lives. Even so, mishaps occur, accidents happen, and birds sometimes need medical attention. Call your area's local vets to learn who is qualified and willing to treat your chickens in the event of illness or injury. Be warned that, despite the growing popularity of chicken keeping, some vets still consider chickens "exotic" pets and may have a higher per visit cost than for a cat or dog. Ask questions and establish these relationships before committing to the purchase of chickens.

2. **Build the Coop.** It may sound overly cautious and a bit silly to have your chicken's coop completely built, painted, bedded, and ready for their arrival, even before your chicks have hatched, but trust me when I say baby chicks grow up FAST. In a few short weeks of brooding chicks indoors, they'll be fully feathered poop machines that you'll be eager to put outside in their own enclosure. Don't scramble to build a coop or order one at the last minute. Like a mama nesting for her new babe, have everything ready before the birds come home.

3. **Find and Meet Other Chicken Keepers.** By no means is this a necessity for keeping chickens, but it certainly is fun to get to know others doing the same thing as you. Practically speaking, establishing a community of chicken keepers gives you valuable (and experienced) eyes and ears to help troubleshoot issues, offer advice and recipes, give tried-and-true solutions to common problems, and much more. Seek out fellow small-scale chicken keepers, rather than large-scale farmers, since these individuals will likely share common goals and have the practical experience that will be most valuable to you. This community need not be physical; in fact, there are fabulous online communities and forums dedicated to keeping chickens. BackYardChickens.com is a personal favorite. If you don't find one you like, or want to start a physical club with a virtual connection, social media sites such as Facebook and Meetup.com have user-friendly online tools to get you started. My local chicken club members post pictures of their birds, swap hatching eggs, find homes for unwanted roosters, and even help each other clean their coops.

do not conform to their breed's standard. Either way, they'll likely make great pets and backyard egg layers, as long as they're healthy, of course.

Taking Over Another Backyard Flock. For someone looking to get out of keeping chickens, you may be their saving grace. In fact, taking on a whole flock from another backyard keeper might be the best situation for you, too. Ask around locally and check sites like Craigslist.org for individuals rehoming their flocks. Remember to ask the usual questions about age, breed, origin of the birds, and general health. Also, ask if the birds have been handled and if they are familiar with families and children (if that's important to you). You may end up with a healthy, relatively young flock that is docile in nature and will meet your egg laying needs nicely.

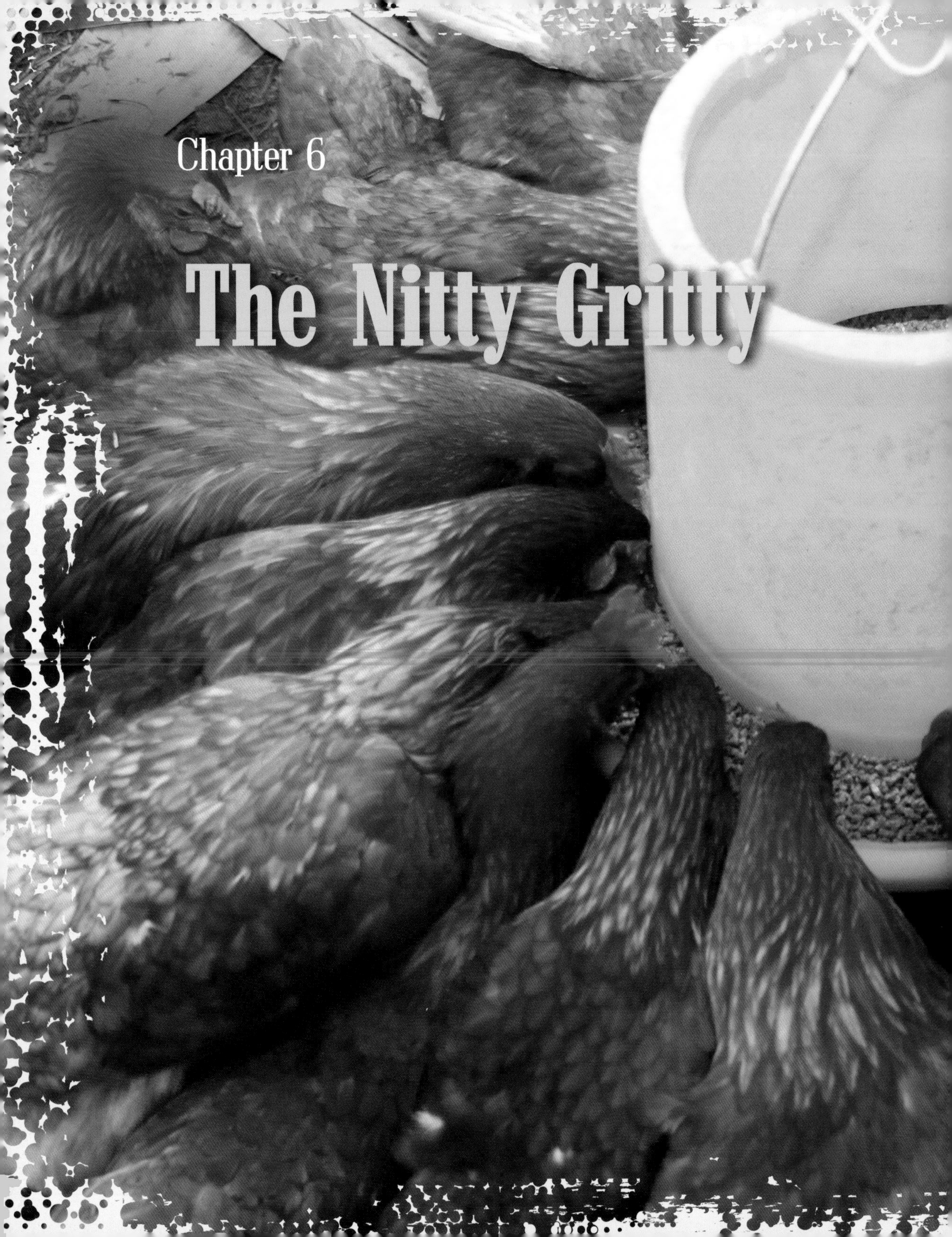

Chapter 6

The Nitty Gritty

Chickens need stuff. They don't need a lot of it, but they need the right kind of stuff to live comfortably and for you to care for them properly. There are many areas in the world of keeping chickens where you can either spend a pretty penny or save time and money through creative means. Sourcing housing and equipment for your birds is one of those areas. In this chapter, we'll explore the basics of poultry housing, from the highest roosts down to the lowest bedding and every nut and bolt in between. We'll also look at equipment to feed, water, roost, nest, hatch, and brood your birds to help make chicken keeping easy and fun. In each section, you'll learn about the various equipment options, their cost ranges, which items you absolutely need, which you can live without, and which items you can make yourself.

Most of this chicken "stuff"—the ready-made, finished product or the material with which to build it—isn't too hard to find either. Keeping backyard chickens is becoming more and more popular in more and more cities, so you'll likely be able to source much of this equipment at your local feed and seed store, pet store, or farm and garden store. Small shops that focus on a variety of homesteading endeavors are also popping up everywhere. When in doubt, absolutely everything you could ever want or need is available online (often with free shipping options), so go to "Resources" on page 214 for a list of some tried-and-true online retailers.

Housing

Where your chickens live and spend most of their days (and nights) will be one of your top considerations as a new chicken keeper. A proper coop protects your flock from the elements and predators and provides a safe place for your hens to lay their eggs. In short, a good coop should meet their every need.

The Ideal Chicken Coop

You can spend as little or as much as you want when setting up poultry housing. Some manufactured coops begin in the hundreds of dollars and range into the thousands. Moveable coops are generally less expensive to buy than stationary coops, and, of course, the larger the coop, the larger the cost. However, this is a great area to reuse and recycle. An unused outbuilding, shed, doghouse, or even a child's playhouse can be easily retrofitted as a chicken coop by adding the features listed throughout this chapter.

Whichever route you decide to take, don't skimp on your coop. Build or buy a sturdy structure that will last a long time, is easy to clean, and something you can be proud to show off and enjoy looking at. You (and your neighbors) will be glad you did.

Whether you build or buy, there are several factors you should carefully consider when planning your coop.

Size

A general rule of thumb when it comes to poultry housing size is that "bigger is better." If your space and budget allow, build the largest coop you can manage and afford. Simply put, the more space your birds have, the happier they'll be. Chickens love to wander as they forage and range, so a spacious coop will allow them to focus on (and enjoy) their daily behaviors of scratching, pecking, dust-bathing, and preening. A coop that is too small can result in birds that feel very confined and resort to cannibalism to relieve boredom and stress.

But don't make it big just for the birds. You'll want to build a spacious coop for your needs, too. Any veteran chicken keeper will tell you that keeping chickens is an addicting hobby. After a few years, you may want to try your hand at hatching your own eggs or raising a rare breed, for instance. However it happens, you'll likely want to add to your flock down the road, and if you build the coop big enough from the get-go, you'll be comfortable in the knowledge that everyone will have ample living space. Plus, a smartly built coop is easy to clean, and if the coop is easy to clean, then you'll clean it!

This petite henhouse sits in a small garden, allowing the chickens to scratch, peck, and dust-bathe close to home. Free-range time is a must in a coop of this size.

So how big should this coop be exactly? There are many guidelines, but it varies by a keeper's personal preference, the breed of chicken you keep, and even the personality of the flock. Generally speaking, 3 to 4 sq. ft. (0.3 to 0.4 sq. m) of space per bird is a minimum of space for standard-sized birds *if* they are allowed to free-range during the day (meaning they are able to spend that time outside of the coop). Allow about 10 sq. ft. (0.9 sq. m) per bird if they are confined to the enclosure full-time or most of the time. A happy medium is

A Guide to Chicken Housing

Chickens don't need much to keep them happy and healthy. Below is a description of the basic items you will need to get you started.

Coop. Any shelter that houses chickens may be considered a coop. Tractors, arks, poultry housing on wheels, stationary poultry housing, or converted barns or sheds are all examples of coops (as long as chickens live in them, of course). The ideal chicken coop is large enough for the number of birds it houses, offers protection from the elements (of all seasons) and predators, has adequate ventilation, is free of drafts, provides the hens a place to lay, and is easy to clean. With access to grass and/or a large outdoor run, a coop needs only to provide space to lay eggs (see "Nest Box" opposite), roost at night, and hang water fonts and feeders.

Tractor (or "Ark"). A chicken tractor is a small, lightweight, floorless, moveable structure that houses chickens, usually on a temporary basis. The tractor's flexibility is its draw. The design allows gardeners and farmers to move chickens to empty garden beds, overgrown pasture, compost piles, and other locations that they want gently tilled and fertilized. However, if confined to one space for a long period of time, chickens will destroy all vegetation and leave only droppings and dry dirt behind (hence the name tractor). For this reason, tractors are intended to be moved often, even daily. Some designs are light enough that one or two people can move them; others are on wheels for easier portability.

Run. An outdoor enclosure or pen attached to a coop is called a run. Most runs are stationary, are covered to protect the birds from the elements, and have flooring material, such as sand, to aid with drainage. Creating a large, outdoor run attached to a smaller coop is a great way to meet the minimum space requirements for your flock without breaking the bank on a large coop.

Pen. A pen is a fenced outdoor area in which chickens can forage and range. Pens are usually moveable and may or may not have a covering. They can be large or small. Lightweight, electric net fencing creates a spacious, predator-secure pen in which chickens can forage during the day. Metal dog or puppy training pens work well as a temporary run for chicks.

These hens are content, moving in and out of their solidly built coop. Make your coop as simple or extravagant as you want.

This hen looks content laying eggs in her nest box. Hens like to lay eggs in dark, dry, secure places.

Pop Door. This is the name given to the chicken-sized door of the coop. A pop door should be just large enough for a chicken to walk through comfortably, be able to close flush with the wall of the coop, and lock securely.

Nest Box. The nest box is a five-sided box lined with soft bedding where hens are encouraged to lay their eggs. The ideal nest box is dark, dry, soft, and secure. The perfect nest box is located far from loud noises and intruders and is positioned away from very high traffic areas. (Laying eggs is hard work, and your ladies will appreciate the privacy.) Some nest boxes are wall-mounted in the interior of the coop; others are designed to be accessible from the outside of the coop, making daily egg collecting supereasy.

Roost (or "Perch"). A chicken's roost is a bar, pole, rough dowel, or piece of lumber mimicking a tree branch in size and texture. This is the preferred place to sleep for domestic chickens. Roosts should be placed a minimum of several feet off the ground and ideally higher than the coop's nest boxes. Allow plenty of roost space—at least 10 in. (25 cm) per chicken, more if space allows.

Chicken Wire. Also called hexagonal netting or hexagonal wire, chicken wire is a thin mesh made of twisted wires that create its distinctive honeycomb pattern. Like other livestock mesh, it's available in various size openings. Unfortunately, this material is unsuitably named: The mesh is actually too flimsy to provide proper protection for a flock of chickens and is easily torn by determined predators. While I wouldn't recommend this material for the lining of windows, doors, or runs, chicken wire does have its place; it can protect a flock from aerial predators in outdoor enclosures or create dividers between birds in indoor housing.

Hardware Cloth. This material is what good housing and predator protection is made from. Like chicken wire, hardware cloth is a wire mesh that is available in a variety of sizes. Unlike chicken wire, the grid of galvanized-steel squares that make up hardware cloth creates a sturdy (but workable) mesh that provides ample predator protection. While significantly more expensive, a secure coop and run made with hardware cloth is worth the investment.

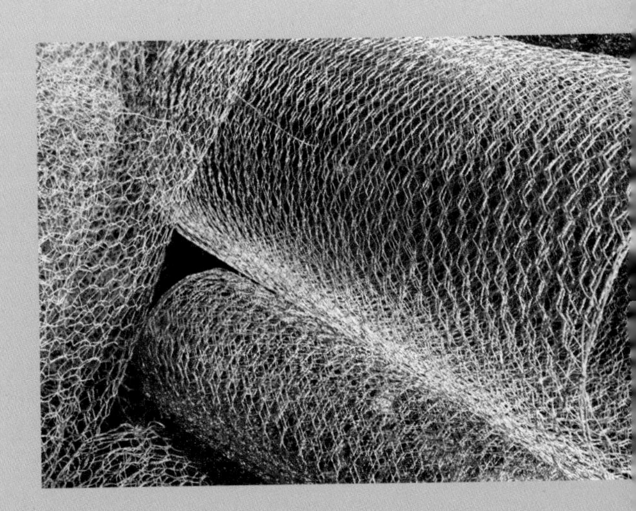

Chicken wire can be used to keep chickens confined to an area but won't keep predators out. Choose a sturdier material such as hardware cloth to secure your coop.

a spacious coop and covered run that you feel comfortable confining the flock to during inclement weather, while allowing a bit of free-range time each day. Since there are no hard-and-fast rules, err on the side of more if you're able. Ample room in the coop is truly the first step to keeping a healthy flock of chickens. Flocks that don't feel "cooped up," as it were, lay more regularly, are less likely to cannibalize eggs or each other, and have stronger immune systems, leading to the ability to naturally fight off pests and diseases.

Location

Your town, city, or suburb will likely have regulations regarding where a chicken coop may be placed on your property, including the minimum distance from existing structures. First and foremost, follow these guidelines to stay in compliance with your city's ordinance. Otherwise, don't get too hung up on the location of your coop. Chickens do fare far better in cold than in heat, so choose a shady spot over direct sun if you have options. If full sun is all that is available to you, just make sure the flock has protection from the heat of the sun with a reliable roof on the coop and run and, of course, access to fresh water at all times (which they should have anyway).

Another consideration for your coop's location is drainage. Moisture is not a friend of the chicken, so locate your henhouse out of areas that tend to get muddy or flood during heavy rains. Use a porous and easily drying material (more on that in the "Bedding" section on page 97) on the ground in outdoor runs to allow for quick evaporation after rains.

Don't be afraid to locate your coop close to your home. After all, it will need to be easily accessible for feeding, watering, and daily egg collecting, rain or shine. Even a well-maintained coop will have an earthy chicken smell, but it should never smell *bad*. And, with the exception of the daily "hen song" after laying an egg, a flock of hens is very quiet, so your coop can be as close to or as far from your home as you prefer.

Materials

From plastic to cedar, prefabricated chicken coops are available in practically any exterior-grade material. While wood is the most traditional coop material, it's not the only one that will do the job. When choosing a material for the coop, shop for something sturdy that is easy to maintain and clean.

When choosing an interior material, opt for something free of chemicals or treatments. If you use paint on any part of the coop, run, or nest boxes, keep it on the outside only. Remember that chickens investigate the world with their beaks, and everything inside a coop is fair game to peck (and presumably eat). Otherwise, your ladies won't be picky: As long as the coop does its job, your flock won't care what it's made from.

Ventilation

An often overlooked component of coop design, proper ventilation in any poultry housing is absolutely paramount. At the risk of sounding like a broken record, moisture and chickens simply do not mix. Soggy or damp housing environments can create habitats for unwelcome guests, such as mites, parasites, and disease. Chickens, like most birds, have very sensitive respiratory systems, making good airflow critical for good respiratory health.

With that said, care must be taken not to inadvertently expose your birds to cold drafts. What is the difference between *ventilation* and *drafts?* Ventilation allows air to circulate and moisture to evaporate, without allowing cold winds to blow onto roosting or nesting birds. Drafts occur when large areas of the coop remain open, allowing gusts of chilly wind to enter, particularly around roosts at night. Windows are a great source of ventilation since they are easy to open during the day and close at night if the temperature drops, but they shouldn't be the only source. Provide vents near the eaves of the coop's roof or on either end of the coop.

Protection from the Elements

Have you ever seen a wet chicken? Aside from being rather pitiful, chickens become pretty ornery when they're uncomfortable. And who could blame them? They're just not made to get drenched (or overheated, or snowed on, or blown around). Simply put, chickens don't fare well in extreme weather conditions. That includes deep snows, heavy hail and rain, and the beating sun, to name a few. While generally pretty scrappy, exposure to some of these extremes could leave your birds with serious injuries that may even be fatal. So, one of the coop's main functions is to protect your flock from these natural elements, particularly at night when they are the most vulnerable.

A good coop provides daily access to shade, no matter the season. It offers a dry area to roost, take cover from rain, and a dry place to lay eggs year-round. In cold and snowy climates, a good coop provides a windbreak from the harsh winter winds and a covered area free from deep snows that could contribute to frostbitten toes and feet.

Protection from Predators

Protection from predators is arguably the most critical element that a coop provides but one that is sometimes the most difficult to achieve—just ask any chicken keeper who has lost birds to a fox, weasel, raccoon, or other predator. Even with savvy, hungry

Tales from the Coop

Those of you reading this in northern, wintry climates may be wondering about heating your coop. This subject matter is sorely contended among many backyard chicken keepers, and the jury has always been—and will likely remain—out on this one.

Most keepers fall into one of two categories, those who believe their chickens need a leg up in the cold and those who feel the risks of coop heating outweigh the benefits. While I firmly believe that we are completely responsible for our birds' well-being, I personally fall into the latter camp.

Chickens that are allowed to acclimate to the cold gradually, over the course of the fall and early winter, are able to fare better through the ranging temperatures of winter than those who are given supplemental heat. With adequate ventilation, the use of the Deep Litter Method (see page 97), quality food, fresh water, and overall good health going into winter, chickens fare rather well in cooler temperatures—certainly better than they do in heat.

Finally, it's important to note the inherent risks of artificially heating the coop. Unless a generator or off-grid power source backs up your coop's electricity, an outage could plunge your flock's temperature from something rather cozy to something rather deadly. Such a drop in temperature could spell disaster for the entire flock in just one night. What's more, the combination of livestock heating apparatuses and dry, dusty bedding equal a very realistic fire hazard. Those that advocate against heating the coop often cite the many chicken keepers who have lost flocks to fire rather than the cold they were trying to protect them from.

Make sure your coop is locked every night. Avoid latches like this carabiner that, even though it looks sturdy, can allow a persistent predator to open a locked door.

prowlers lurking, protection from predators should be at the top of every chicken keeper's priority list. How you outfit and secure your coop will depend on the type of housing you use and the most common predatory animals that roam your area. Here are a few coop security hot spots that should be on your radar.

Locked Up: Pop Door, Doors, and Windows. Stationary coops will need a locking door (also called a pop door) that is closed every night. For this door's lock, choose a fastener that requires the use of your thumb and forefinger, such as a locking bolt latch or a spring action hook-and-eye lock with a spring-loaded safety latch. (For under $5 each, a good lock may be the cheapest security measure you take in the chicken yard.) Simple sliding bolts and hook-and-eye locks (that do *not* lock) are too flimsy and may be easily popped or jiggled open by persistent and dexterous predators such as raccoons. Also, try to avoid using latches such as carabiners that, due to their large size, allow the locked door to open slightly, such as when being pushed or pulled.

Some chicken keepers have perfected the art of the automatic coop door system, where a mechanized pop door is set to a timer and automatically closes at a certain time each day. When done right, this is a great system. If you choose to install a similar system, remember a few things. First, teach young pullets where home is and make sure that all birds know where their coop is and are retiring to it each night at a consistent hour *before* you set the timer. Also remember that the time of sunset changes throughout the year and by region. This gradually shifts the time that a flock will retire for the evening—earlier in the winter and later in the summer—so adjust your timer accordingly.

Windows are a great source of airflow and sunlight in any coop, but they also tend to be a security weak point. Make sure the windows in your coop are able to close securely and use the same type of locks mentioned above. Screen all windows with a sturdy mesh, such as hardware cloth, for predator protection when they're left open.

Mind the Gap. When shopping for or building a coop, look at every nook and cranny from a predator's point of view. Are there any weak points where a predator may gain access? What are they and how can you fix them? Either reinforce the structure or pass on purchasing that coop.

The roof, windows, and all doors of a coop should be secure and flush with the walls of the structure. Any gap, even as small as a single inch, may be large enough to let some predators into the coop (think snakes, mice, and weasels) and give others just enough access to do damage (think raccoons). Windows should be securely stapled with hardware cloth mesh, and roofs should be leak-proof and free of gaps with all siding secure and flush to the walls.

You may be reading this section and thinking to yourself that these security measures sound a bit over the top and overprotective. And on paper they may seem that way. But life in the coop is a different world. Remember that chickens are the quintessential prey and have very few natural defenses. Pretty much everything wants to eat them, and even those that don't, such as domestic dogs, still see them as prey that are fun to chase, bite, maim, or kill. As the guardian of the flock, it's your responsibility to keep them safe. Provide them with a coop that protects against all predators, large and small. It will help you—and your birds—sleep better at night.

Flooring

The type of flooring you choose for your coop will largely depend on its style, size, and function.

Tractors, arks, and other moveable pens are, by design, floorless and *open*—the "flooring" will be whatever ground you move it onto, allowing the birds access to the earth, grass, and bugs below, fertilizing as they go. Some chicken keepers line their tractor's bottom with a large-weave wire mesh to keep digging predators from getting in while still giving the birds access to the ground to forage and fertilize. This feature is only necessary if the birds live in the tractor full-time or if they aren't locked up in it each night. While a wire bottom is a savvy feature in terms of predator protection, it actually prevents the birds from comfortably scratching as they forage, since their footpads and toenails hit wire each time they try. If you use this tractor design, routinely check for sharp spots in the wire mesh and monitor your birds for foot injuries on a regular basis.

Stationary and semi-stationary coops have several flooring options. Many prefab chicken coops are constructed on legs with *plywood* flooring. This raised coop/flooring design is desirable for various reasons: It keeps a variety of predators out, it isn't apt to flood, and it's easier on the chicken keeper's back come cleaning time. The flooring of the roosting and nesting space in tractors and arks is usually made of plywood as well. Lengthen the life of your coop's plywood flooring by lining it with a sheet of vinyl or linoleum before spreading the coop's litter: This will protect it from moisture and be much easier to clean than bare plywood.

If you find a suitable location for your coop that you are very happy with, you may opt to pour a *concrete* floor for it. In terms of predator protection, concrete is an ironclad flooring option. Unless those raccoons come armed with jackhammers, they likely won't be getting in through the floor.

Tractor Tip

Even a small flock of chickens can reduce a patch of lush grass to barren dirt in a day or two of normal scratching and pecking, so whether your chicken tractor has a wire bottom or not, you'll want to move it daily to keep the flock from decimating a single spot in your yard.

Cedar Shavings–To Use or Not to Use

Pleasant aromas are a rare topic in the world of chicken keeping, making the use of cedar shavings as bedding a bit controversial among keepers. Cedar shavings are very similar to pine shavings in shape, texture, and availability for use as bedding (although they are slightly more expensive). Cedar is often used as an alternative to pine, but there is much debate as to the strong aromatic nature of cedar shavings and its effect on chickens' delicate respiratory systems. Some chicken keepers believe the difference (and thus toxicity) lies in the type of cedar shaving used—Western red cedar or Eastern red or white cedar, with the Western variety being the most potent. The wood shavings leach plicatic acid, or aromatic oils, that may cause respiratory and skin irritation, among other effects. The smelly-good hydrocarbons, or phenols, that are emitted by cedar wood are the reason it effectively repels some insects and inhibits the growth of some microorganisms. But for some, the risks outweigh the benefits.

Currently, there is no research on exactly how it affects animals of the avian variety, so when it comes to using cedar shavings as bedding in your coop, the choice is up to you. To play it safe, I'd recommend avoiding the use of cedar shavings with very young chicks confined to a brooder full-time. Chicks tend to pick at their bedding and may eat the shavings. Many chicken keepers have used cedar shavings with success in their coops with adult flocks, as long as the birds have other areas to inhabit besides the coop and the enclosure bedded with cedar shavings is very well ventilated.

If you're wary of using cedar shavings in the coop, the nest box is a great place to try it out with your flock. Unless you have a broody hen setting a clutch of eggs for several weeks at a time, the birds are rarely in a nest box long enough for any aromatic oils to disturb their respiratory systems. Watch your flock for signs of change or discomfort and then decide if it's right for your coop.

In terms of flexibility—well, it's not. Concrete is relatively permanent, so make sure you really love the location before you pour.

Some prefab coops designed for rabbits or other small livestock may be marketed toward chicken keepers for the ease of cleaning, thanks to their *wire-bottom* flooring. However, wire-bottom flooring is not an option I'd recommend. Without much padding on their feet (not to mention their lack of feathers or fur), chickens are prone to a host of foot maladies, especially the heavier breeds. Wire-bottom flooring invites injury and discomfort. It also eliminates any chance for normal chicken behavior, such as scratching and dust-bathing. Part of the joy of raising your own chickens is doing so responsibly, knowing they are humanely cared for and that they live dignified lives filled with the activities chickens naturally engage in. The best way to do this is to provide comfortable flooring, plenty of daily access to the outdoors, and opportunities to forage...and pass entirely on the wire flooring.

Finally, some chicken keepers prefer to build their coop without any flooring at all, placing the structure on bare earth. While cheap and easy, *floorless* coops do not have suitable protection against digging predators, such as foxes and dogs. If you choose to go floorless, you'll definitely need to take other precautions against these predators, such as lining the perimeter of the coop with a foot of buried hardware cloth. Floorless coops may still require several inches of bedding to combat moisture and dampness as well.

Bedding

Whatever type of flooring you choose, you'll need to line it with a few inches of bedding. Despite its cozy-sounding name, bedding isn't a luxury. It serves many important purposes in the chicken coop. The right kind of litter provides a secure foundation for a chicken's legs and feet (preventing all sorts of leg ailments and injuries), helps you gather droppings quickly for use in the garden or compost, allows you to implement the Deep Litter Method (see below), provides a soft landing for those precious eggs, and makes your coop cleanup an easier task.

So what's the best stuff for the job?

Sun-Colored Straw. With its sweet, earthy smell and springy texture, sun-colored straw is what many new chicken keepers reach for to line their coop and nest boxes. It may look beautiful, but straw or hay, because it's hollow, retains moisture too well to make it suitable for long-term bedding. Moisture breeds disease, and a bedding of soggy, damp straw is the perfect environment for some pretty nasty poultry pests to thrive. As pretty as it looks, the price to pay in damages isn't worth outfitting your coop with straw. Keep straw for back-up bedding or for temporary housing, such as a broody pen.

Deep Litter Method

The Deep Litter Method (DLM) is a form of coop maintenance that requires a little preparation, some timely attendance, and a bit of calculated laziness. The idea behind the DLM is simple: Begin with several inches of bedding material and build the layers, lasagna-style, over the course of many months. The droppings and bedding will gradually decompose (essentially composting), giving off ambient heat in the process. An additional inch or two of bedding should be added once every month or so, depending on your coop, how many birds you house, and how much time they spend in their coop. Use a rake to periodically turn the litter over, and add diatomaceous earth (DE) or dried herbs to keep mites and lice at bay. Wood shavings are the best bedding material for use in combination with DLM; sand will not break down or provide heat.

DLM requires no more bedding material than any other method or more than you would otherwise be using. But due to the sheer volume of bedding added over time, it does tend to trickle out as the birds enter and exit the coop. Use a plywood board or something similar at the coop door to contain the bedding. If you begin to see flies, pests, or an excess of manure or it begins to smell, it's time to add more bedding (and DE) and be more attentive to turning it over regularly.

At the end of a year, you'll likely have a hefty amount of composted matter to add to the garden or the compost pile. Use your eyes, nose, and good judgment to determine when you need to replace the bedding and start again. A clean, properly cared for coop should never smell bad. The best part about DLM is that it requires little maintenance, infrequent cleanings, and saves the chicken keeper's time, energy, and—most importantly—his or her back.

Pine Shavings. A popular and prudent choice for coop litter—and my personal preference—is pine shavings. Pine shavings dry fast and don't break down quickly, making it an ideal bedding material. It's also easy to remove soiled bits without having to replace it all. The mild pine scent is inviting and non-irritating for your birds, although the scent does fade over time. A bale of pine shavings is inexpensive (usually about $6) and is easy to source since it is a popular choice for poultry and equine enthusiasts. Look for it at feed and seed stores, big box stores, and even pet stores.

Sand. Another great bedding option is sand. When used as coop bedding, sand is an excellent and very clean choice for those who are willing to spend a little more money. The initial investment for sand is a bit more than for wood shavings, but it doesn't break down nearly as quickly, so you won't have to replace it as often. Sand does require some regular cleaning, though: Turn it over once a week with a rake or scoop out the large clumps of droppings to keep it dry (think of it as a really big cat litter box). Sand is also a great material for covered, outdoor runs: It dries quickly, doesn't break down, and, when mixed with diatomaceous earth, doubles as a great material for dust-bathing. In outdoor runs, sand will likely need to be replaced more often than in closed coops, since the birds toss the material out as they dust-bathe. Even so, most chickens love it, as mine do. One of the only true drawbacks to using sand is that it does not retain or give off heat, so it can become cold in the winter. Sand is probably best for summer use or for use in warm climates. If you use sand in the coop, opt for builder's sand over sandbox sand, since the latter is very fine and tends to clump when wet.

Natural, Reclaimed, and Recycled Materials. There are many natural, found, and recycled materials that could make fine bedding in theory—it doesn't necessarily mean they should be used. Some should probably be avoided completely.

Grass clippings are a viable option for bedding, and while free, have a few disadvantages. Clippings tend to retain a bit of moisture and break down quickly. They also dry, shrink, and smell. If you opt for grass clippings in the coop, be sure to stay on top of the decomposition process so you know when to add more bedding or clean it out completely. Most importantly, use only clippings that come from a yard that has not been sprayed with pesticides, fungicides, herbicides, or other chemicals. Chickens will pick at anything and everything in their coop, and bedding is no exception, especially if there may be bugs in it.

Shredded leaves are another natural option for coop bedding, but only if they are finely shredded so they dry quickly. Whole leaves take a very, very long time to break down and are susceptible to harboring moisture, sticking together, and compacting into slick clumps. Wet leaves make a slippery surface that could lead to splayed legs, bumblefoot, or other foot injuries, especially in younger, growing birds.

Pine shavings are a popular and inexpensive choice for bedding material.

Save Bedding: Use a Droppings Board

Chickens make the majority of their waste during the night when roosting, so the best way to be frugal about the bedding you purchase and use in your coops is to install a droppings collection board under the roosts. A droppings board is a board or tray designed to catch the nighttime waste, sparing your bedding from accumulating large amounts of manure. Using droppings boards will also help your coop stay dry and clean—well, as clean as the inside of a chicken coop can be. Droppings boards also make transfer of pure waste to compost piles easy and efficient—no shoveling or sifting through bedding to get to the "garden gold." A droppings board will need to be emptied on a regular basis, as determined by the size of your board and the number of chickens you keep. It's a great method that can be used alone or in combination with the Deep Litter Method (DLM). When used in tandem with DLM, bedding lasts much, much longer.

To make a droppings board, begin with a shallow bin (think of a raised garden bed) made of a material that is easy to lift and clean. The best size is several feet wide and as long as your roosts. If you have very long roosts, make several trays that sit end-to-end along the bottom of the roosts. Next, create a removable lid of heavy-duty wire mesh to cover the top. This will allow the droppings to fall through for collection but prevent the birds from stepping in their droppings.

Finally, there are a host of reclaimed and recycled man-made materials that are potential bedding options, such as shredded newspaper or shredded office paper. These would seem to be great alternatives since they are also free, but use them with caution. Ink can be toxic to chickens, and office paper is heavily processed and treated. Glossy paper, such as the kind found in magazines and fliers, also contains a large amount of ink and can create a matted and slippery surface. Call your local newspaper to get the lowdown on their printing process; some newspapers are printed with soy and water-based inks. These bedding options are safest for coops in which chickens don't spend 100 percent of their time.

Roosts

It's a common misconception that chickens spend their nights sleeping soundly in the confinement of their nest boxes. But as ground fowl, domestic chickens learned that their survival hinged on being able to roost and sleep as high as possible, out of the grasp of anything that goes bump in the night.

The ideal roost is simple to find and even easier to make. A 2 x 4 piece of lumber, sanded smooth and turned on its short side, makes a perfect roost. The semiflattened edge of the lumber allows a bird's feet to extend comfortably, making it possible for her to spread her feathers over her legs, feet, and toes and keep them warm. The texture of unpainted, untreated, sanded lumber is also just right, making it easy for your chickens to grip and roost comfortably all night long.

This hen sits comfortably on her roost.

If you want to get creative, you may find sturdy tree limbs or branches that make suitable roosts. An ideal branch is 1½ to 2 in. (4 to 5 cm) in diameter, free of thorns or sharp branches, and very solid. Try not to use anything made of plastic or that is too smooth, such as a closet dowel; the rounded shape and smooth finish make it difficult for chickens to balance comfortably and rest well.

Finally, the perfect perch is a spacious one. Allow about 10 in. (25 cm) of roost space per standard-sized bird and a bit less for bantams, about 8 in. (20 cm). As with coop square footage, when in doubt, provide more space. If the birds get chilly during the winter nights, they'll huddle closer together for warmth and spread out in the heat of summer.

Nesting and Nest Boxes

A laying hen, whether she's hatching her eggs or not, has particular needs and preferences when she searches for a nest. You'll find that some hens are really picky and others will lay in the oddest places—under a roost, out in the open, even in your flower pots. Sometimes a single nest box will be so popular among the ladies that the "it" girls (really the alphas) will go first, and the rest of the flock will line up behind them, patiently waiting their turn to lay in that box.

Whether broody or just laying her daily egg, each bird has her own nesting pattern and laying routine and will do anything to find a great place to lay. From the hen's perspective, each egg is a potential chick that must be given the best chance for survival. The hen is a resourceful creature and will create a nest she loves if she can't find one to her liking. So, as guardian of the flock, your responsibility isn't to please everyone, it's to give your hens a place to lay that will appeal to their predilection for a nest. Soft, dark, dry, and safe are the key elements for a suitable chicken nest.

Soft. A soft nest is important for a hen's level of comfort and for the safety of her eggs. Soft nests serve your interests, too. If you are keeping laying hens for their eggs, you don't want to find them broken when you come around to collect. A hen will naturally want to

Of Roosts and Nests

To keep birds from roosting—and therefore defecating—in their nest boxes, keep the boxes at or close to ground level. Many chicken keepers with high wall-mounted nest boxes (the kind the hens have to fly up to) find that some birds mistake the perches used to enter the nest boxes as roosts. Injured birds, broody hens, new layers, or birds low on the pecking order are candidates for this confusion, seeking safety away from bullying by sleeping on the nest box perches rather than on the roosts with the rest of the flock. If bullying is a concern, check your coop's square footage to be sure they have adequate space. Then, if you keep nest boxes and roosts separate and distinct, you'll have clean eggs, a tidy coop, and happy hens.

lay in a spot that protects her delicate eggs from breaking. The same materials for bedding in the coop (with a depth of at least 4 in./10 cm) make a suitable lining for a nest. My preferred materials are pine shavings, since they do not retain moisture the way others do and are much easier to clean when removing soiled bedding or broken eggs.

Dark. Most hens prefer a dark and secluded place to lay. Again, we can chalk it up to instinct: The safer a hen's nest and the more protected it is from predators, the higher chance of survival for her eggs and subsequent chicks. Make sure nests are in a dark place, securely mounted and/or stable in their location. If the nest is very bright, consider using some old sheets to create a "curtain" of fabric in front of the nest's entrance. This also helps prevent egg-eating behaviors by keeping the freshly laid eggs out of sight and out of mind of naughty hens.

Dry. Finally, keep the nest boxes dry. Eggs are laid with a protective outer coating called a *bloom* or *cuticle*. This coating protects the developing chick from potentially harmful bacteria (and if your eggs are unfertilized and just for eating, it keeps the egg fresher longer). Most hens instinctively know that water destroys the bloom, which may put the chick at risk for illness or death. For this reason, hens will look for a nest that is out of the elements. Secure any leaks in the coop and be sure to provide proper roofing on the nest boxes to keep conditions dry for them.

Shape and Size

Young pullets are malleable and impressionable creatures and can easily be taught to lay in just about anything, as long as it meets their needs (listed above). Just about any five-sided boxlike object will do. You may choose to purchase premade nests or create your own. Cardboard boxes may work for a time but will become soiled and may harbor mites or bacteria and become moldy and smelly. Old milk crates work well, as do small pet carriers or even covered cat litter boxes. Being plastic, all three of these options would be rather easy to clean. Building nest boxes out of wood is simple and inexpensive, and you can line them with large plastic containers for easy cleaning as well. Get creative and don't be afraid to recycle. It's more important that you choose a box that is easy to clean over one that looks fancy.

Lighting the Way

When the days begin to shorten and daylight turns to dusk earlier and earlier, your hens will gradually slow down their egg production. Come winter, it may appear to cease altogether. This is nature's way of cueing the hen to push the pause button on reproduction. Nature knows that frigid temperatures are not conducive to raising heat-loving chicks, so a hen's body stops producing eggs, whether they are fertilized or not. (Add the annual molt into the mix and you may be looking at several long weeks, even months, without consistent fresh eggs.)

While this makes sense for the chicken, it's a bit of a bummer for the chicken keeper. When you've had fresh eggs all summer long, it's hard to give them up come winter. But some thrifty chicken keepers find a way around this dilemma. They know that a hen needs 14 to 16 hours of light exposure per day to lay a single egg. When the days shorten, it takes longer to get this minimal exposure, and thus egg production is delayed (so it doesn't really stop, per se, it just slows down).

How does the chicken keeper get around this natural cycle? He or she will artificially light the coop. Using the same red poultry bulbs used for chicks in a brooder, the keeper can extend the daylight hours by simply turning on a light in the coop. If you employ this method, it's wise to put these winter lights on a timer to extend the day just enough to trigger egg laying, while still giving the flock enough darkness at night to rest well.

There's one caveat: Artificially lighting the coop is. . . well, artificial. Among natural chicken keepers, this practice isn't always condoned. Like the gardener or farmer who has worked hard all growing season, many believe that laying hens deserve a bit of a break, too. If you choose to light the coop, keep an eye on your flock. Watch for signs of picking or pecking, increased stress, or cannibalism. Whatever you decide to do, use your best judgment while considering your region's daylight hours, your management practices, and your flock.

If you are purchasing a coop that has nest boxes already built in, make sure you check that the size of the box is appropriate to your birds' size. Bantams require a nest box that is at least 10 in. (25 cm) wide x 12 in. (30 cm) high x 10 in. (25 cm) deep. Standard-sized birds need a nest box about 14 in. (36 cm) wide x 14 in. (36 cm) high x 14 in. (36 cm) deep and a few inches larger for extra-large hens, such as Jersey Giants. Both sizes will accommodate a few inches of bedding material, too. The boxes should be just large enough for a hen to turn around in and sit comfortably—think cozy and snug but not tight. Finally, the roofs of nest boxes should be sloped to discourage birds from perching above the boxes and soiling the eggs or the laying hens below. Only one nest box is needed for every four to five hens in a flock.

Location, Location, Location

As we know, our domesticated chickens like to roost high and nest low. Hens are resilient, though, and most will adapt to nest boxes placed high, such as those mounted on the wall of a coop. Young pullets just learning to lay, on the other hand, will instinctively try to create their nests closer to the ground. They'll just need some patient redirection in the beginning.

Leave new nest boxes at ground level for several weeks until your young pullets are familiar with laying in them. When they are laying regularly, you may choose to mount the boxes. If you do so, fasten them no higher than 3 ft. (1 m) from ground level and always provide a perch in front of mounted nests for hens to land on when flying up to them.

If you notice a pullet laying her eggs on the floor of the coop or in the old location of the nest boxes, gently place her in the new nest box every time you see her sitting. You may have to be persistent

Diatomaceous Earth

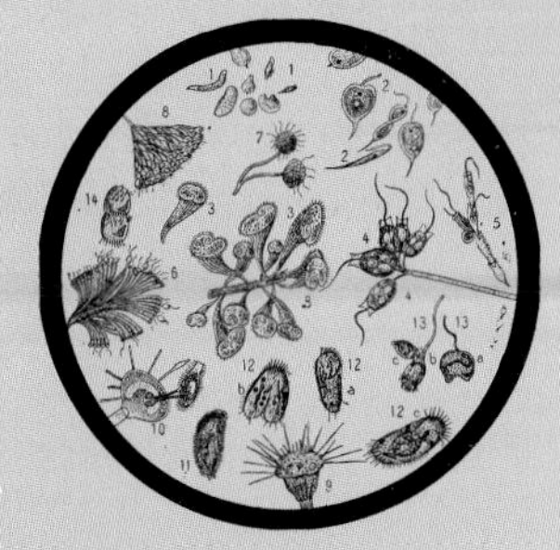

Formed millennia ago from the fossilized remains of tiny aquatic organisms called diatoms (shown here), diatomaceous earth is often used as a natural insecticide in the chicken coop.

Diatomaceous earth (pronounced die-uh-tuh-MEY-shuhs) is a mineral-based substance made from the fossilized remains of diatoms, or algaelike water plants. Its uses have been widespread since its discovery in 1839, from an absorbent and stabilizer in dynamite to a filtration medium for swimming pools. Many chicken keepers have employed diatomaceous earth (known simply as DE for short) as a natural insecticide in the chicken coop. The common understanding is that the jagged edges of the DE particles pierce the waxy exoskeleton of common pests that come into contact with it, killing the insect through dehydration. It also has the potential to work as an effective nonchemical pest repellent as bugs learn to stay away from the area where it is present.

DE looks and feels like powdered sugar. While it is rather deadly to insects, it is soft to the human hand. But DE is also a very abrasive respiratory irritant, due in large part to its high silica content. There are no regulations or recommendations on its use, and some research has even pointed to a possible increased rate of lung cancer and respiratory diseases among workers in the DE industry. Knowing how sensitive the chicken's respiratory system is, it would be prudent to exercise caution when using DE in the chicken coop.

While DE is indeed a natural substance (in its pure, 100 percent food grade form only, of course), we now know that not everything that is "natural" is good for us, for those we care for, or for the environment. Harvesting DE requires extensive strip mining, a process that destroys wildlife habitats, leads to soil erosion, increases the risk of chemical contamination, pollutes waterways, and poses other environmental risks. The environmental impacts of strip mining are vast. For this reason, it's important to carefully evaluate the use of DE in your own coop and consider whether this is an industry you wish to support.

Furthermore, it's important to consider DE's impact on your backyard environment once in use in your home. While it may be a non-chemical "pesticide," it does not discriminate between "bad" bugs and "good" bugs. In other words, the use of DE has the potential to harm beneficial insects and pollinators, such as honeybees and native bees, in addition to the pests you are trying to manage in your coop, such as mites and lice.

What is the solution, then? Admittedly, I have used DE in my coop with good results. But I also employ preventive management with my flock. They are fed high-quality feed, have fresh water daily, and their coop, run, and surrounding environment are kept clean. I occasionally boost their health with supplements and herbs, when needed. So is it the DE that is effective or my management practices?

My recommendation is to exercise stringent biosecurity measures and practice the management techniques recommended in this book. If you do choose to use DE, purchase only 100 percent food grade, which is often found at feed stores or homesteading shops. Be wary of DE sold at large home improvement stores and read the ingredient lists. Contact the manufacturer if you are unsure about anything regarding the product—there should be nothing in it but DE. Completely avoid DE products made for gardens and swimming pools, which contain harmful additives.

Some chicken keepers use diatomaceous earth to repel common pests, such as lice and mites.

with physically moving a new layer for a few days. With consistency, she should adapt to the new location within a week or two. Generally speaking, hens will only remain in a nest box for the time it takes to lay an egg—about 30 minutes per day, once a day and usually in the morning. If you see one sitting for a very long time or overnight, chances are she's broody, injured, or ill. Many birds retreat to the safety and seclusion of a nest box if they aren't feeling well. Inspect her more closely and refer to chapter 9 for troubleshooting illnesses.

Alternative Housing

Whether your flock lives in a small moveable tractor or in a 200 sq. ft. (19 sq. m) veritable hen palace, it's wise to have back-up housing in case a hen becomes ill, is injured, goes broody, or otherwise needs to be isolated from the rest of the flock. Back-up housing may seem prudent, but over the life of your flock, this need will arise more often than you think. Chickens are prone to cannibalism in certain circumstances. Isolating a sick or injured chicken from her flock mates is often necessary to prevent it.

Temporary housing need not be fancy—it's temporary, after all. It should provide the basics: safety from predators and the elements, something to roost on, and a nest box. It should also hold a small feeder and water font that are easily accessible to the isolated bird. (This is a great reason to hold onto that chick-rearing equipment: Spare chick-sized feeders and water fonts are perfect for a single bird.)

Any traditional housing will work well: A small tractor or ark is a perfect back-up coop. A medium- to large-sized pet crate placed in one corner of the coop also works well to keep sick or injured chickens separate from her flock mates and still within the safety of the larger coop. In a pinch, even a cardboard box will suffice for short-term temporary housing, such as monitoring an injured chicken indoors overnight. The most important thing is that you have a plan in place for housing when a bird needs it. If your budget or space constraints don't allow you to keep spare housing around, at least consider your options in advance. What will you do when a bird falls ill? Search online for free or cheap poultry housing for sale in your area. Also, take inventory of your home's spare rooms where you might temporarily house a chicken (garages, basements, and mudrooms are all viable options). Finally, get to know other chicken keepers or farmers in your area. Your poultry-passionate community will offer a wealth of information and support, and some may even be able to rehab injured chickens for you.

Equipment

Once housing is checked off the list, there are a few sundries that will make feeding, watering, and otherwise caring for your flock much simpler. Many a chicken keeper has creatively made water fonts and feeders from reclaimed objects or easily (and cheaply) sourced materials. However, my suggestion to new chicken keepers is to create space in your budget to purchase ready-made feeding and watering equipment, either new or used. Beginners need some time to get to know their birds and their habits before being able to design equipment for them, so in this case, leave it to the poultry professionals. Luckily,

this is one area where purchasing commercially made equipment doesn't break the bank. Let's start with the most critical element to chicken health: water.

Waterers and Water Fonts

The most popular way to water a flock of chickens is with a galvanized steel or plastic water font. These cylindrical buckets are gravity-fed and utilize natural water pressure to keep from spilling. A lip around the bottom offers a continuous flow of water to the birds, and it gradually refills as they drink. Water fonts are available in sizes that hold as little as 1 qt. (1 L) of water (typically used for chicks or individual adult birds) to several gallons. Many come with handles and/or hardware for hanging, which is a smart feature that prevents birds from roosting on the waterer and soiling the contents. This popular water font design is widely available and tends to do its job very well.

Other watering options include chicken fountains (made of PVC pipe), automatic fountains that continuously cycle water, and nipple waterers made from large buckets or bottles. There are even brightly glazed ceramic waterers and feeders available for purchase that make fancy garden accessories. Finally, while chickens will drink out of a water bowl or basin if offered, these containers can be stepped in or knocked over and are easily soiled, so only use them in a pinch.

Feeders

The many varieties of commercial poultry feeders are as easy to source as water fonts. They often look quite similar to poultry water fonts, since they are made in the same cylindrical shape with hanging options. It's also pretty easy to find font/feeder sets in coordinating size, style, and color.

Feeders work in much the same way as water fonts. Feed is poured through the open top and filters down through holes around the base, filling the lip. As chickens eat, more feed filters down. Most plastic feeders have "teeth" or tongs spaced out every inch or so along the base to prevent the birds from scratching or tossing the feed out with their beaks. It's a smart system and works well when it's set up properly.

Troughs, Basins, and Racks

Every chicken flock needs a few supplements to stay healthy. These supplements (think grit and oyster shells) should be

Metal feeders like these are sturdy, hold a lot of feed, and work well. The feeder is filled from the top, and as the chickens eat, feed filters down through holes around the base, filling the lip.

Fun Fact

Ever wonder why all the commercial chicken feeders are red? Chickens are drawn to the color red, and poultry keepers realized they could train chickens to eat from feeders more easily if they incorporated the bold color.

Feeding/Watering Tips

The Right Height. Whether you hang your feeders and fonts or place them on the ground, make sure they are stationed at the back height of the chicken. This is generally a comfortable height for chickens to eat and drink, and it also prevents them from stepping in the water or scratching out the feed with either their feet or their beaks. Scratching and picking for food is a natural chicken behavior, after all, but it can become quite annoying when feed is continually wasted. Give them time to pasture in addition to a filled feeder, and you'll have happy hens and a full wallet.

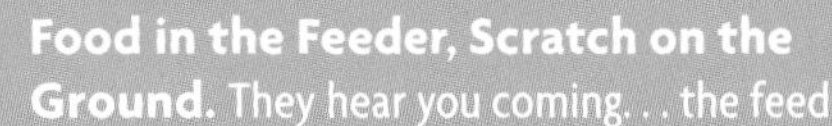

Food in the Feeder, Scratch on the Ground. They hear you coming. . . the feed clatters against the pail, rattling with each of your footsteps. They know what's in the bucket and eagerly gather around as you scatter the feed to the ground. It's fun to watch your birds flock to your side in their half-run, half-waddle, and there's something very gratifying about tossing feed and watching them gobble it up. That's part of the fun of keeping chickens, after all. Unfortunately, throwing feed to the ground is not a very efficient way to feed a flock. In the long run, this method wastes valuable food and takes a serious toll on your wallet. My advice? Stick to keeping your flock's layer ration in a designated feeder. Less than $20 will buy you a feeder large enough for a medium-sized backyard flock, and you'll likely not have to make that purchase again. And when you want to give your birds a treat, try scratch. It's fun and easy to scatter by the handful, and the morsels are large enough that they won't be lost in the grass or bedding. It's fun for kids to do, too.

housed and offered separately from the feed. One option is to use a very small chick feeder for each of the supplements. Another option is to fill a small, heavy-duty plastic livestock trough. Farm supply stores often sell two-chambered troughs that are easy to hang along the side of the coop or run. Either vessel is perfect for offering supplements, and both make it easy to see when your supply is low. As with the feeder and water font, keep all containers holding supplements at back height.

For various treats and kitchen scraps, shallow galvanized steel bins or rubber livestock basins are perfect. They're easy to clean out and stay put while the chickens scramble for the treats. Wide plastic bins with low sides are great for offering treats and are always easy to clean, but chickens tend to get excited when there is food involved: They'll probably end up flipping over the lightweight containers, spilling the contents.

"Snack racks" are another great tool for offering treats to your flock. Heads of lettuce, cabbage, and other greens work well in these racks, which offer tension for the birds to pick at and grab small, digestible pieces. Snack racks are easy to find at pet stores, labeled as hay dispensers for rabbits,

hamsters, pet rats, and similar animals, or you can make one from leftover mesh or chicken wire. Just be sure the material isn't easily broken and that all sharp edges are secured.

Miscellaneous Feeding Equipment

There are a few simple items you will want to keep on hand to make feeding more efficient.

Grain and Feed Scoop. A grain scoop is by no means a necessary piece of equipment, but having one certainly does make scooping feed easier. At just a few dollars, this item doesn't break the bank either. Feed scoops are commercially available in a variety of sizes, from just a few cups up to 6 qt. (6 L), and a variety of materials, such as plastic, wood, and galvanized steel. Steel scoops are the most popular due to their larger size capacity (2, 4, or 6 qt./2, 4, or 6 L) and their ease of use and ease of cleaning.

Feed Bucket. Unless your feed is stored in your coop or immediately adjacent to your bird's feeder, you'll probably need a bucket to schlep it from the feed bag to the feeder. Buckets made of stainless steel or plastic are both wonderful easy-to-clean options, but just about anything dry will work. Look for one with a comfortable handle, too. As for size, shop for a bucket that holds about the same capacity as your flock's feeder. This coordination makes it easy to fill the feeder each time and also to gauge how much they need or when they will run out.

Feed Storage Containers. Whether you have a microflock of three or four chickens or a budding poultry operation, you'll need to store a good amount of chicken feed for several weeks or months at a time. Most commercial chicken feed comes in fairly large bags (50 lb./23 kg bags are standard), although urban homesteading and some feed and seed stores may offer food and other supplements (such as grit, oyster shells, and scratch) by the pound. This gives you the option to purchase as little or as much as you want at any time.

Store your chicken's feed in a secure container that keeps out moisture and pests. Just about any airtight food-grade storage container will do. Mice, rats, raccoons, and other critters will find chicken feed rather tempting, so either use a container that locks or store feed in the coop behind a closed door. Food-grade plastic buckets and pet food storage containers make great options for small amounts of feed. For larger quantities, my personal preference is a large galvanized-steel trash can, which can fit two 50 lb. (23 kg) bags of feed perfectly, and the lids are easy to secure. Keep containers raised just above the ground (such as on blocks or bricks) to encourage airflow.

If you're building your own coop (and it's large enough), it's helpful to build in a storage area for equipment and feed. It's pretty clear that chickens need some stuff, so storage should be a consideration when you're establishing your flock. When building or creating a storage space, consider how far it is from the coop. You will want to have easy access both to the coop and your storage area (and the space in between), rain or shine.

Wire egg baskets are sturdy and easily found at antique stores, online, or through homesteading supply catalogs.

Egg Equipment

There's nothing like grabbing the egg basket and heading out to the coop to see what the ladies have been up to. Keep a small basket and clean egg cartons on hand to keep eggs from banging together.

Egg Basket. With just a few hens and a small handful of eggs daily, it may be tempting to completely forego the basket and simply carry the bounty back to the kitchen by hand. However, homegrown eggs are precious, not to mention fragile, and they deserve the best treatment. A designated egg basket is especially important if you have a family flock where children are tasked with chores, such as egg collection. Using an egg basket helps reduce breakage, of course, and helps children stay organized and focused on the task at hand.

Nearly any small basket will work well to collect eggs: a small market basket, a berry basket, or a vinyl-lined wire egg basket, made just for this task. For children, choose something easy to carry, preferably with a handle. Even an egg carton will do in a pinch and makes easy work of putting the eggs straight into the refrigerator.

Egg Cartons and Miscellaneous. The thriftiest way to stock up on egg cartons (aside from saving your own) is to ask friends and family to keep their cartons and give them to you. If you plan on selling your eggs at a tailgate or farmer's market, you'll want something you can personalize with your farm or family's name on it. In this case, there are many styles of blank egg cartons available through retailers that are neutral and ready for personalization. These retailers also offer printing services, stamps, and stickers to indicate the size of the eggs and details about their source, such as "pastured," "free-range" or "organic," and your farm's name.

False Eggs. False eggs are placed in nest boxes to help young pullets or new birds learn where to lay. They also help discourage egg eating by teaching young birds that those round things in the nest boxes are hard on the beak, inedible, and not worth their time (of course we know that's not true, but that's what we want them to think!). Use false eggs mindfully, however: If left in nest boxes all the time, these decoys may inadvertently trigger a particularly motherly hen's broody instinct.

Poultry suppliers make lovely false eggs made of smoothly sanded wood (and they're generally not very pricey), but any object that is generally egg shaped will do. Golf balls and plastic Easter eggs are also inexpensive and easy to clean to boot.

Chick Equipment

If you plan on raising your flock from chicks, you'll need a few additional items.

Brooders. Think of the brooder as a mini-coop or "crib": It's a soft, safe place for young chicks to grow. A brooder provides safety and security and mostly keeps them contained. Brooders are not meant to stand alone outside or weather the elements like a coop, but rather, a brooder provides a contained area for the chicks to remain warm and grow in the safety of a larger shelter, such as a barn, shed, big coop, garage, or basement. As

mentioned earlier, you don't need to purchase a commercial brooder for rearing chicks: A kiddie pool, bathtub, or even a cardboard box will suffice, as long as it keeps the peeps contained and safe and it is placed in an area free from drafts.

Heating Lamps, Bulbs, and Thermometers. The best method to heat a brooder is by using a poultry heat lamp with a 250-watt bulb. Even the new models sold today are rather simplistic. A heat lamp turns on by plugging it in, and it has only one setting: on. To turn it off, simply unplug it. Coupled with an infrared 250-watt heat bulb and suspended above the brooder (either hung or clipped), a single brooder lamp will warm small clutches of chicks nicely. When shopping for a heat lamp, you may notice styles with and without a guard on the brooder: *Always* purchase one with a guard. Should the lamp fall into the brooder, the bulb will be protected from breakage and prevent the lamp from coming into contact with the bedding, which could be a potential fire hazard. Some poultry supply catalogs also sell lamp stands for just this purpose.

Your lamp will hold many different bulbs, but the best bulb to buy is one designed for producing heat for poultry. Poultry supply catalogs will likely give you options between red and white bulbs: Always buy the red. White bulbs produce harsh light that doesn't give the chicks a chance to rest, inevitably leading to cannibalism and general stress. Red bulbs soften the light, allowing the birds to go about their business in peace. Remember, the lamp will be on 24/7 for up to a month to provide warmth—that's a long time in the life of a chick.

Finally, invest in a chick-safe thermometer. Brooder temperatures can't be left to chance or "feel." It's best to have a concrete number of the brooder temperature at various times of day. Thankfully, poultry thermometers are very inexpensive and are easy to find among chick-rearing supplies.

Hatching Equipment

When you've been keeping chickens a while and you're ready to hatch your own, you have two options: Allow a broody hen to sit on a clutch of eggs, hatch them, and raise the chicks for you, or you can hatch them in an incubator.

An incubator is an artificial mother hen. It is a piece of equipment designed to mimic the temperature and humidity of a broody hen and the duration of time that she sits on a nest of eggs.

There are dozens of styles and models of incubators available, from mini-hobbyist models (that hold about 10 to 20 eggs) to industrial-sized incubators (that may hold over 500). Some require you to manually turn the eggs, while others are automatic. Most hobbyist incubators range between $80 and $350; large cabinet incubators go into the thousands of dollars. While not a necessary piece of equipment by any means, an incubator is a fun addition for those with some chicken-keeping experience under their belts, who want to experience the joy of watching their eggs hatch. Great for 4-H and school science projects, kids, especially, find the incubating and hatching process fascinating.

Incubating your own eggs can be an educational, fun, and rewarding experience.

Chapter 7

You Are What You Eat

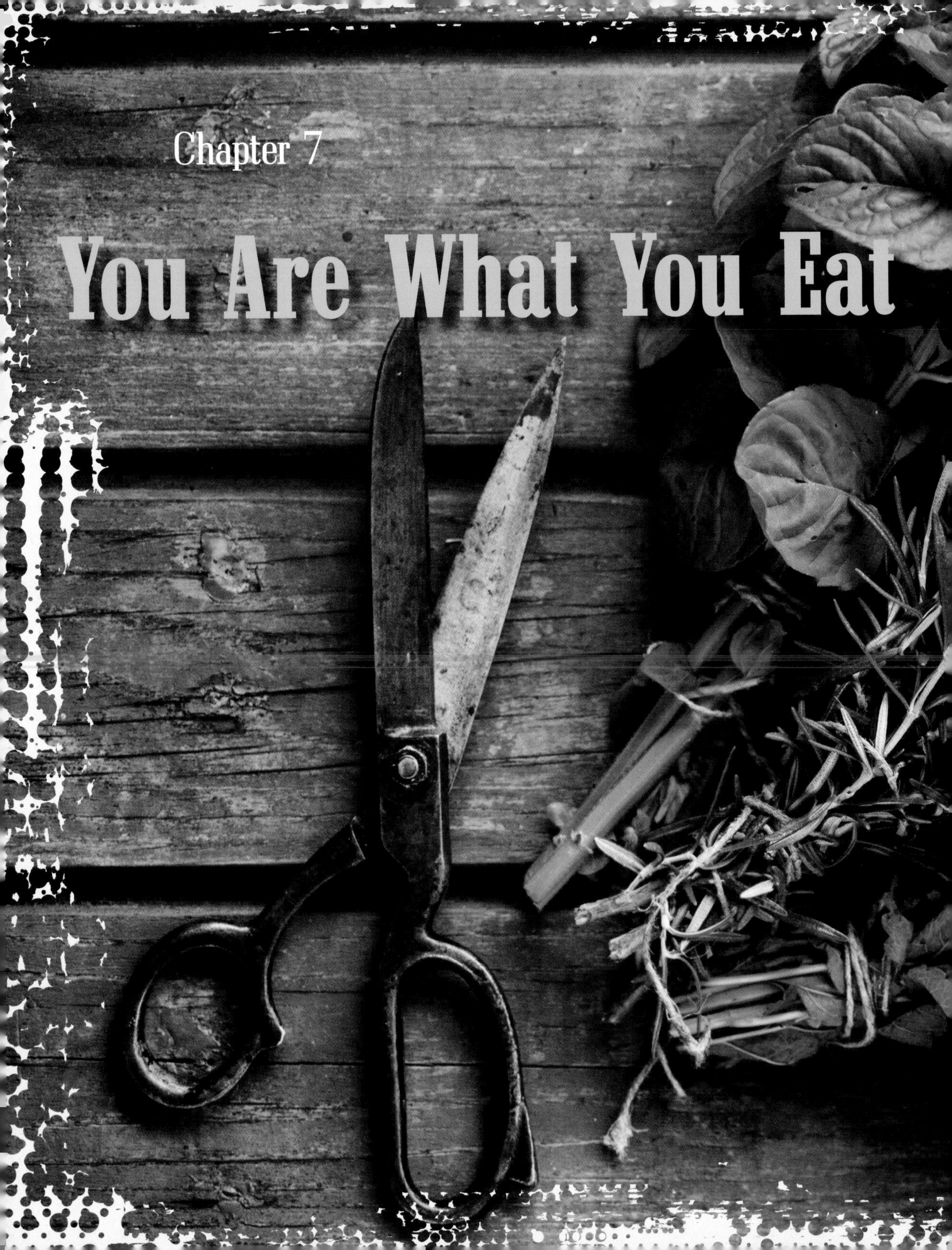

Once you learn what to feed them, how often to feed them, and you have a good feeding and storage system established, the process of feeding your chickens will become rather effortless. It won't take long to get into an easy rhythm of filling the feed bucket, then moseying out to the coop, filling the feeder, and collecting the eggs. In fact, I find the routine rather meditative in its simplicity—there's something really gratifying about bringing nourishment to your "ladies," watching them eat so excitedly, and then gathering a handful of eggs in return.

As with most growing animals, a flock of laying hens requires different nutrients at different life stages and throughout different seasons of the year. And your charges will source it from the selection of balanced feed you provide. This chapter covers the basics of flock nutrition, the feed you'll need, and what to look for in a quality commercial feed. Your flock's extra dietary and digestive needs will largely be met by the variety of supplements provided in addition to daily feed—we'll cover those in more detail, too. Finally, nothing wins over a flock of chickens like a good treat. My advice to any new chicken keeper is to provide a balanced daily feed ration first, allow some forage time, and then shamelessly enjoy bonding and building trust with your chickens by offering fun treats (in moderation, of course). That's the secret every chicken keeper knows: The way to a chicken's heart is through her crop. Let's look at how to get there.

Chickens are opportunistic and will forage for a variety of edible plants, seeds, and bugs.

For Starters: Forage

Eating—that's what chickens do best. Nature has expertly designed them to convert a variety of edible plants, insects, seeds, and other living things into fuel for their daily activities: dust-bathing, grooming, roosting, egg laying, chick rearing, and evading predators (staying alive is hard work, after all). And chickens are certainly happier when they're able to realize their true nature. Foraging is a very basic part of being a chicken. It's their livelihood.

So why have modern chickens been converted to a diet consisting almost solely of cheap grains? Well, unlike ruminants such as cattle (whose seven stomachs evolved to digest grasses best), chickens are fairly good at converting grains and legumes, such as corn and soy, into the protein of their meat or eggs. But just because they *can* doesn't mean they *should*. When lacking a varied diet (and in particular, the greens found on pasture),

Green Grows the Garden

If your flock does not have access to pasture, for whatever reason, a great way to give them their greens is by growing a small chicken garden. Really, a chicken garden is no different from your own. If you already grow food, they'll happily gobble up your garden scraps and extra greens. But if you find yourself really limited on space, try these options for growing greens for your gals.

Container Gardening. Salad greens (like mesclun mixes and arugula); dark, leafy vegetables (such as kale, spinach, and chard); cherry tomatoes; herbs; and so many more cultivars will thrive in pots. Place them on a sunny windowsill indoors, plant them in window boxes on your house or coop, or find a corner of the flower bed to put in a few edible plants. Your birds won't need much, but the variety to their diet will vastly improve their health and the nutritional value of their eggs.

Sprouting. It's so easy to create an indoor sprout "garden." Moreover, it's inexpensive, available to just about everyone, and the health benefits are enormous. All edible seeds, grains, and legumes may be sprouted. Wheat (preferably organic), corn, barley, alfalfa, mung beans, peas, pumpkin seeds, radishes, and carrots are all popular choices for sprouting.

The Chicken-Proof Garden. Many chicken keepers like to grow a mini-garden for their girls but keep it in the coop to save space. Won't they destroy it before it has a chance to grow, you say? The way around this age-old dilemma is to create a barrier between the chickens and the soil so they are unable to scratch and dig. The best way to do this is by creating a shallow garden bed with a wire mesh covering. The wire protects the soil from the chickens while allowing the greens to grow up and through the mesh. Once the greens are through the mesh, the chickens can eat the tops but leave the root-ball intact and growing, providing a continuous supply of healthy greens.

a chicken's egg and meat quality plummets; the good omega-3 fatty acids decline, and the omega-6 fatty acids skyrocket. It's no wonder the chicken egg has developed a reputation for being an unhealthy source of cholesterol and "bad" fats and is often written off as a harmful food. In reality, the modern chicken has just been producing the best egg it can on the feed it is given.

Upon learning this, my knee-jerk reaction was to set all of my hens free to forage for their food on our property. But after learning more about chicken nutrition, I realized that this extreme wasn't the best route, either. Unlike their ancient Asiatic ground fowl ancestors (or the feral Key West chicken, for that matter), domesticated poultry are rarely able to meet all of their nutritional needs by foraging. Small backyards or suburban lots do not have substantial fare to support a flock for very long, if at all.

The bottom line is that variety—and balance—is best. In addition to a well-rounded commercial feed, give your gals a chance to stretch their wings and legs by grazing on grass for a few hours each day. Just because they eat commercial feed ration doesn't mean they need to stay cooped up; any time spent foraging will be great for morale, alleviating boredom, providing exercise, and supplementing their diet. They'll be happier if they can realize their true "chicken nature," and both you and your flock will find the time fun and entertaining. And, if your birds must stay confined to a coop, run, or enclosure, give them some greens as treats—better yet, grow them. (See "Green Grows the Garden" above for some great ideas.) Your chickens, their eggs, and ultimately your diet will be healthier for it.

Chicken feed formulas vary widely, so be sure to read labels and check ingredients before purchasing.

The Main Course: Feed

Like other prepackaged animal and pet food, commercial poultry feed is a relatively new commodity. Before the advent of manufactured feed and industrial trucking routes, domestic poultry likely relied on foraging close to home: "hunting" for bugs, grubs, and other small insects; eating scraps from the kitchen; and foraging for seeds, grains, and of course, grasses. As omnivores who spend nearly every waking minute focused on finding food, this lifestyle suited chickens well for quite a while. But, over time, domestic chickens were bred toward the most efficient conversion ratios, meaning they were able to convert the least amount of food into the largest amount of protein (whether that be eggs or meat). In the last century or so, intensive breeding has turned the domestic chicken into a veritable egg-making machine. This means that the modern layer needs much more in the way of nutrients than her ancestors in order to stay healthy and keep up with her body's egg-laying demands. In combination with healthy greens or pasture time, a good commercial feed is crucial to keeping each hen in her best health.

What's in Chicken Feed?

Each brand of chicken feed will have a slightly different "recipe." The best way to know exactly what is in your bird's food is to take a glance at the label.

Corn and soy are two foods often found at the very top of the ingredients list on a bag of commercial chicken feed, with wheat or wheat meal often a close third. Aside from being high-allergen foods, corn and soy are nearly all genetically modified organisms (GMOs) unless purchased organic. Furthermore, cross-contamination of fields is leading to adulterated wheat as well. The good news is that there are more and more organic choices readily available, and some brands are now making soy-free and even gluten-free chicken feeds for keepers with intolerances or for those who opt to stay away from GMOs. (See "You are What Your Chickens Eat" on page 117 for more on this topic.)

Other ingredients in commercial chicken feed may include oats, field peas, flaxseed, rice bran, fish meal, oils (such as sunflower), alfalfa meal, kelp, and a variety of yeasts and

added vitamins. But because brands vary widely in their ingredients, additives, and preservatives (if any), read labels closely and call the manufacturer if you have questions. Chicken feed should be consumed within about three months from time of purchase; even though it will still be edible, its nutritional value rapidly declines beyond that time.

Crumbles, Pellets, or Mash?

Chicken feed is available in three types, or textures: crumbles, pellets, and mash.

Crumbles. With their, well, *crumbly* texture, crumbles are the most popular form of commercial chicken feed. Crumbles are actually just crushed pellets. Despite its popularity, crumbled feed may lead to some wasted food because it spills rather easily from the feeder.

Pellets. Compact capsules of feed, pellets are specially designed to contain a balanced ratio of nutrients in each morsel. In fact, this is the major advantage to feeding pellets to your birds: They receive complete nutrition in each pellet; no picking out the pieces they like and leaving the rest. A major disadvantage is that birds tend to eat pellets rather quickly, become satiated, and then bored without anything to pick and eat—and bored chickens will usually pick on each other instead. A flock with a lot of pasture time will do well on a diet of pellets.

Mash. Very simply, mash is all of the individual ingredients of a chicken feed recipe mixed together but easily recognizable to the birds. Chickens fed mash tend to pick out what they like and leave the rest. This is a major disadvantage to feeding mash. Much of the feed is wasted if the birds refuse to eat certain ingredients. Plus, ignoring some of the less desirable ingredients may lead to

Feeding pellets to your chickens offers a balanced ratio of nutrients, since your hens won't be able to pick out just the pieces they like and leave the rest.

One of the major disadvantages of feeding your birds mash is that much of the feed is wasted if your hens refuse to eat certain ingredients. One way around this is to mix the remaining feed thoroughly with water-your birds will love it.

We've all heard the admonition *Don't count your chickens before they hatch.* The ancient Greek fabulist Aesop is credited with coining a version of this phrase, but what exactly does it mean? For many, it acts as a gentle warning against making plans or assumptions based on something that is unpredictable. For the chicken keeper, however, incubating eggs—a very delicate and tricky endeavor resulting in about a 50 percent hatch rate—it can be taken quite literally!

nutritional deficiencies. There are ways around this, however. The easiest is to make a wet mash with the remaining feed by adding water and mixing thoroughly—the birds love it. The biggest advantage to feeding mash is that, unlike pellets that have been processed into morsels, the ingredients in mash have never been heated or processed and retain many of their nutrients.

Feed Rations

Whatever texture or brand of feed you choose to buy, it's most important that you purchase the appropriate feed for the age and life stage of your flock. The feed should also meet the hen's needs based on the season and temperature and her breed's weight, size, and rate of lay. Generally speaking, chicks and breeder birds require a bit more protein, layers require a bit more calcium, and almost all birds eat a bit more of everything in winter and a bit less in summer. Of course, this will vary from region to region and bird to bird.

When to graduate from chick to adult feed and how to make the transition will largely depend on the brand you choose. Read the label thoroughly and call the company to learn how and when they recommend transitioning your birds from one feed to the next.

Starter. Starter feed is designed for newly hatched chicks. Starter ration may be fed for just the first two to three weeks of life, or, until they are fully feathered and reach point of lay, depending on the brand and its formula. Chicks grow incredibly fast and have some significant nutritional needs while they're growing. They're working hard to put on weight and keep up with developmental needs, not to mention, they're growing an entire body's worth of feathers in the process. These needs account for the average 20 percent protein in most commercial starter feeds. Additionally, starter feed is available in medicated and non-medicated versions. The medication added is a *coccidiostat* that protects against the disease coccidiosis and is an alternative to vaccination. Choose either the coccidiosis vaccination or the medicated feed, but not both. Given together, the two will cancel each other out.

Grower (or Developer). This ration is designed for birds past the chick stage but still growing. Most chicken keepers feel comfortable feeding this ration to their birds until they reach point of lay, and then they switch to layer. Some brands entirely omit this type of ration and create a "starter-grower" combination feed that is suitable from hatching until point of lay. With either recipe, transition the flock from grower to layer ration when the pullets reach point of lay (more on that below).

Broiler and Finisher. You likely won't be using either of these rations if you're raising a flock of laying hens. *Broiler* feed is for meat birds that require a substantial amount of protein (about 30 percent) when growing (which is something they do rather rapidly). There are different types of broiler feed designed for meat birds at different stages of life. Also, depending on the type of bird being raised for meat—hybrids, heritage, or free-range meat birds—and their stage of growth, the protein requirements will change slightly. *Finisher* ration is designed for meat birds in the weeks before processing.

Layer. Layer feed is perfectly calculated for the nutritional needs of egg-laying breeds. It boasts a lower protein count than most of the other types of feed at 16 to 18 percent and has the additional calcium and minerals that most laying hens need to remain healthy

You Are What Your Chickens Eat

Have you ever heard the phrase "You are what you eat"? Well, it's no different when it comes to raising chickens for eggs. The quality of feed you provide them will directly affect the health and nutrition of their eggs—which will directly affect your health and nutrition as you eat those eggs. For instance, many people with gluten or soy intolerances show sensitivities to factory-farmed eggs but are able to eat eggs from pastured hens. This is because the factory-farmed hens may be fed products that contain these and other potential allergens. Pastured hens, on the other hand, live on a diet that largely consists of natural grasses, seeds, and bugs and will be supplemented with feed (hopefully organic). If you or a family member experience any sensitivities to certain foods, especially top allergens like soy, consider purchasing chicken feed that does not contain these ingredients and provide even more pasture and free-range time to your flock.

You may also want to consider purchasing organic chicken feed. Organic feed may come with a slightly higher price tag, but it also comes with better overall results and health benefits. Some of the food products that are most likely genetically modified organisms (corn, soy, and sometimes wheat), or GMOs, also happen to be main ingredients in nearly all commercial chicken feed. The only way to be sure that you aren't buying or eating GMOs is if you purchase certified organic products. If you do purchase feed that contains corn, cornmeal, soy, or soybean meal, consider buying organic to reduce your, your family's, and your chickens' intake of GMOs. After all, you're raising chickens to feed yourself and your family the best food possible—and it starts with what your chickens eat.

To DIY or Not? Making Your Own Chicken Feed

We chicken keepers are an interesting bunch. We're returning to our roots and taking our food choices into our own hands. We're taking a front-row seat to see how our food is made and playing an active role in how it gets to our plate.

So it only seems natural that backyard chicken keepers would want to make their own chicken feed. I get it—I've been there. But the truth of the matter is that homemade chicken feed has more drawbacks than many know. Before you make your own feed, consider these points:

- Homemade does not equal "cheaper." In fact, when you calculate the cost of individual ingredients (which are apt to fluctuate), mill costs, your time, your transportation, and even the bags in which to store the feed, a small backyard chicken keeper will rarely reap a cost savings going the homemade route.
- Unless you're already an expert in the area of avian nutrition, creating a formula of chicken feed is a complicated matter. Most backyard keepers simply lack the complex understanding of chicken nutrition that is required to create feed suitable for their bird's nutritional needs. Feed companies hire experts to help devise and test their "recipes"—the combination of ingredients and ratios is professionally calculated.
- Working with mills on a small scale is easier said than done. Many mills require you to produce a recipe or use one of their own to produce a custom order. Also consider that mills will rarely produce feed in small batches, leaving you to sell or use the bulk feed before it goes bad. If they do sell custom feeds in small batches, it may be at an additional fee, cutting out any savings you may have scored. Finally, since feed loses nutrition in just a few months, storing large amounts of chicken feed is rarely an option for backyard keepers (unless you're getting into the feed business).

If DIY everything is a personal passion and mission of yours, and your heart is set on homemade chicken feed, contact local grain growers, mills, and employ the services of an animal nutritionist to point you in the right direction. Your county extension office may also be of assistance especially if you live in a large egg- or chicken-producing state. It will take many hours of sweat, blood, tears, and research, but it can be done.

while keeping up with their body's egg-laying demands. Many layer feeds also have roughly 2 percent fat and 8 percent fiber. While it varies from brand to brand, layer is generally best started when pullets reach maturity or point of lay. Depending on breed, this may be between 19 and 25 weeks, although it averages 22 weeks of age. One word of caution with this ration: Do not feed layer to a flock before the birds reach this point in maturity. The higher calcium content of layer ration may cause serious and irreparable kidney damage to young, growing birds.

On the Side: Supplements

There are two main supplements that every chicken keeper should become intimately familiar with: grit and oyster shells. Both supplements look like tiny rocks or very small gravel and are easy to source since they are very common supplements for backyard chicken keepers. Every egg-laying flock will need both of these supplements.

Some chicken keepers will argue that layer feed completely covers the digestive and reproductive needs of laying hens (it can). Similarly, others will insist that by free-ranging, chickens are able to pick up all the "grit" they need from the ground (sometimes they do). But pet chickens are often fed scraps and treats that require additional digestive effort, and pastured birds will eventually run out of natural grit wherever they roam. However you choose to feed your chickens, they will need access to the grit and oyster shells that you provide.

Grit

From beak to vent, the entire digestive system of a healthy chicken is a well-oiled machine. As omnivores, chickens are able to eat practically any organic material, and nearly every bit of it is put to good nutritional use. Very little of what a chicken eats goes to waste, and even then, their waste is considered valuable fertilizer for home gardeners and farmers.

The Why. If there were a cornerstone in the digestive system of a chicken, it would be grit. Without teeth to chew their food, chickens rely on these little rocks to team up with the strong muscles of the gizzard to mash and grind everything they eat into manageable, digestible bits. A bird deprived of grit may experience a variety of digestive issues, such as impacted crop, sour crop, other digestive blockages, and their repercussions, ranging in severity from severe discomfort to death.

It's a common misconception that pastured chickens don't need grit. It's true that they are able to pick up small rocks to aid in digestion as they graze throughout the day. However, some regions (such as those with clay soil, for instance) and small lots or backyards may not have enough natural grit to supply an entire flock of hens for very long—or at all.

Supplements such as grit and oyster shells are necessary for proper egg laying and overall health.

Another common conception is that birds fed exclusively commercial feed do not need grit, either. This is true *if* the birds *never* eat anything other than their feed ration—no treats, no bugs, no grass, no pasture time. Since this management style closely resembles that of factory farm practices, it's unlikely that most backyard chicken keepers will adopt this practice. Since your birds will likely be a family flock or pets, they'll get the occasional treat or forage time, and thus, they need grit.

The What. In the chicken world, grit is defined as small pebbles or rocks that aid in digestion. Grit isn't fancy—it's just tiny rocks, after all—but it works hard and gets the job done.

The When. Begin offering grit to chicks as young as one to two weeks old if they are eating any food other than their designated starter feed (such as treats) or if they spend any time outside on pasture. Choose chick-sized grit for young birds. If this product isn't available at your farm and garden store, check pet stores for parakeet grit or grit for other small pet birds. Once your young flock reaches its adult size, switch to the standard-sized chicken grit.

The Where and the How. Offer grit free choice at all times in its own designated trough. This is best placed where the birds spend the majority of their days and have easy access. Chickens instinctively know when they need more grit in their gizzard and will take it at their leisure. If they don't need it, they won't take it. For this reason, resist the temptation to toss grit to the ground or mix it with anything else, including feed. Contrary to good intentions, this practice won't encourage your birds to eat more. In fact, it's a surefire way to waste a lot of grit. Fill the trough with grit and monitor for a dip in supply. Then refill as necessary. That's it. This area of chicken keeping is an excellent place to trust in

Chicken Obesity: Fact or Fiction?

Another debate among chicken keepers is whether or not chickens can become overweight. We all know that too much in the way of fatty, greasy, sugary, or starchy foods isn't good for anyone. But can a chicken get fat? The answer is yes and no.

The digestive system of a chicken is incredibly resourceful. The chicken's body is designed to turn all manner of food into fuel or eggs rather efficiently. When fed an appropriate, age-suitable feed and allowed to forage (with the occasional treat from her keepers), a chicken will stay healthy, and her feed conversion ratio will remain balanced.

But that's not to say a chicken can't get fat. While an obese bird will not necessarily appear to gain weight in pounds or ounces, she can accumulate an excess amount of fat internally that can lead to health complications. Too much in the way of treats, fatty foods, or even limited access to daily exercise and room to roam can lead to fat building up around the liver and the reproductive system. This buildup of fat can lead to a host of complications, such as reproductive stress, overheating, an enlarged liver, blood clots, hemorrhaging, and sudden death. And to boot, chicken death as a result of obesity is difficult to diagnose without performing an autopsy following the bird's demise; in this case, prevention is the only cure.

So what's a chicken keeper to do? Well, it's surprisingly difficult to gauge when a chicken becomes overweight. As mentioned above, they tend to hold their fat internally, making it difficult to see with the naked eye. They simply don't show weight gain the way mammals do (with surplus weight and skin rolls, for instance). The best way to keep your flock slim and trim is to offer appropriate feed, forage, and exercise time along with important supplements, and to limit fatty treats until after they have eaten their feed.

Finally, a well-balanced commercial feed should be available to your flock at all times. Contrary to popular myth, chickens will not overindulge in their standard feed. In fact, having it available at all times will allow them to graze and eat as they are hungry (mimicking the foraging behavior chickens prefer) and avoid gorging on food in one sitting, especially when you come by with the treat bucket. Offer treats in moderation and not when they are very hungry. Finally, limit treats given in the summer (when chickens should be their slimmest) and save fatty treats, like scratch, for fall and winter, when the flock is trying to bulk up for the cold.

your birds' instincts and allow them to meet their own digestive needs. Your job is simply to make sure the grit is there when they need it.

Oyster Shells

Many of the egg-laying breeds of chickens we hold dear have been developed and refined over time to produce the maximum number of eggs possible. The production of eggs, however, requires a substantial amount of calcium—about 2 grams per egg, in fact. While a laying hen's body is capable of high egg production, her reproductive system may not always be able to keep up with the demand of calcium needed to make the egg's shells. Even the best layer feed can lack proper calcium for certain hens at certain times of the year.

Chicken Lore

A popular myth about oyster shells and grit is that they are one and the same—or, at the very least, that they are interchangeable. Neither myth is true. Oyster shell is soluble, so it breaks down in the digestive tract slowly over time, gradually releasing calcium to the hen. Grit does not dissolve. In fact, grit's purpose is to remain intact and serve as the hen's teeth, breaking down food in the gizzard through grinding.

The Why. What happens when a hen doesn't have all the calcium she needs? Without enough calcium in a layer's diet, she could experience reproductive issues, such as prolapse, and exhibit cannibalistic behaviors, such as egg eating. Her eggs' shells will lose strength over time and may become deformed or even shell-less. Eventually, a hen with a true calcium deficiency will develop weakened and brittle bones as her reproductive system pulls the calcium it needs from her body. Fortunately, the remedy is so simple. Providing a laying flock with oyster shells every day of the year is a quick, easy, and inexpensive preventive measure for a host of reproductive problems—and keeps you flush with eggs, too.

The What. So, are oyster shell supplements actually made from oyster shells? Yes, they are. Commercially available crushed oyster shell supplement is a natural product that looks like small white rocks, about the size of chicken grit, and is a bit chalky. They dissolve in the bird's digestive tract and provide her with the calcium she needs. Because oyster shells are a natural product, manufacturers will usually claim each bag to have a guaranteed calcium analysis of between 33 and 38 percent, rather than a fixed number.

The When. Pullets do not need extra calcium until they are actively laying. In fact, an overabundance of calcium could be harmful to young chicks, so hold off on offering oyster shells to immature birds. A good rule of thumb is to begin offering this supplement to a flock upon maturity (at point of lay), or once the first few eggs start showing up.

The Where and the How. As with grit, a great way to provide oyster shells to your flock is to offer it in a small livestock trough or a chick-sized feeder, hung or placed at back height. Also like grit, oyster shells should be offered free choice at all times. Each bird will take the supplement as she needs it and leave it when she doesn't. A hen will never overdo it eating oyster shells. She instinctively knows when she needs more calcium, and the amount needed will vary from bird to bird. Sometimes, the same amount will sit in the trough for months, and other times, you'll walk out to the coop and notice the flock will have visibly put a dent in the supply—seemingly overnight. Even so, don't fret about refilling the trough often—it's rarely a daily job. A 5 lb. (2½ kg) bag costs less than $10 and will easily last a flock of 10 to 12 hens a year or more. Again, as with grit, simply refill as needed and let the girls do the rest.

Without grit and oyster shells, a flock of laying hens may be healthy for a little while, but a diet lacking both supplements will take its toll on the flock's digestive and reproductive health in time. When both grit and oyster shells are so inexpensive, so easy to find, and offer so much in return, it's hard to make an argument against providing these to your birds. Why take the chance?

Herbs, Flowers, and Superfoods

There are many herbs, plants, and foods that naturally boast immense healing and health-promoting properties. There are herbs and plants that effectively repel nasty bugs (whether fresh in the garden or dried), make excellent tonics and tinctures, have anti-microbial, anti-viral, and anti-parasitic properties, and some that are nutrient-dense foods eaten just as they are. The list of the amazing things these plants and foods can do goes on and on. Your flock may benefit from many of these properties, too. Most of the following herbs and plants are really easy to grow in any size garden or pot, and the so-called superfoods are easy to find at grocery stores and farmers' markets.

Many, aromatic herbs have anti-microbial, anti-viral, and anti-parasitic properties and are effective at repelling bugs in the nest box.

The Basic Herbs: Basil, Mint(s), Oregano, Rosemary, Sage, and Thyme. These highly aromatic and flavorful herbs have spent generations in our kitchens helping to preserve food and adding an impressive array of nutrients to common fare, as well as boosting taste. There's no debating that herbs are powerful to the olfactory sense, which is what makes them flavorful in dishes, and they can be either soothing or abrasive. This is exactly how herbs work at repelling bugs: The strong aromas either confuse or irritate the insect's highly sensitive olfactory sensors, ultimately keeping them away.

What's more, these six potent plants boast anti-viral, anti-parasitic, and anti-microbial properties and are great for use in nest boxes, made into tinctures, or steeped into oils (to then mix with scratch or other treats). Oregano has recently been commended for its powerful anti-microbial properties, boosting the immune system, and helping to ward off colds and flu. The various mints make great rodent repellents when used in the coop. Mint, when combined with rosemary and oregano, makes a great blend for treating respiratory ailments, too. All six of these herbs are exceptionally easy to find and easy to grow, even thriving in small spaces or container gardens.

Calendula, Chickweed, and Clover. All three of these plants make excellent, nutrient-dense foods for chickens. Calendula has anti-microbial properties and can be applied to wounds to speed healing (yarrow is another good herb for wound care). Chickweed and clover are both packed with nutrients and make great tonics or tinctures. These "weeds" are a favorite chicken food and can be pulled straight from your lawn or garden (as long as they haven't been sprayed with chemicals) and given to your birds as they are.

Catnip (Catmint), Feverfew, Lavender, and Lemon Balm. When dried and sprinkled in the nest box and throughout the coop, these four hardworking herbs will use their anti-parasitic properties to keep pests away. Feverfew targets lice and mites, and lemon balm is anti-viral and anti-microbial. Catnip's aroma is particularly relaxing for your birds. Lavender and lemon balm also have similar effects on chickens—their scents are calming and soothing just as they are for humans.

Dandelion and Elderberry (Leaves and Flowers). My yard is sprinkled with dandelions, and it's one of the first things my chickens go for when they have free reign of the yard. It's a good thing, too, since dandelions are a powerful superfood. Elderberry has also received a lot of attention recently for its impressive anti-viral properties. The berries can be made into a syrup or tincture to help reduce fever and keep viruses at bay. Your chickens will find the berries an irresistible treat.

Garlic. Used to flavor foods and keep vampires away for centuries (just kidding!), garlic boasts some pretty powerful anti-microbial and anti-parasitic properties. Though not an herb, the benefits of garlic for poultry can't be overlooked. In fact, if there were only one natural remedy used for poultry, garlic should be it. This powerful plant (of which we usually eat the root, or bulb) is a natural antibiotic, so the body does not build up resistance. Garlic boosts your birds' immunities, fighting all manner of ailments from the inside out.

The key lies in the two main chemical compounds *allicin* and *diallyl sulphides*. Allicin fights bacteria, viruses, molds, and yeasts. Garlic is most powerful as the allicin is breaking down, but before it degrades completely. Highly unstable when exposed, allicin degrades within 24 hours and completely breaks down within 20 minutes at high temperatures, such as when cooking at high heat. Diallyl sulphides, on the other hand, do not degrade quite as quickly. For this reason, crushed, raw garlic is the most powerful source of allicin and the most likely way to effectively transmit its health benefits. The trick is getting your birds to eat it.

Crushed, raw garlic with its powerful anti-microbial and anti-parasitic properties offers huge benefits to your flock. In fact, if there were only one natural remedy used for poultry, garlic should be it.

Essential Oils, Infusions, and Liniments

Essential oils are liquids distilled from the leaf, flower, stem, root, or part of the plant that, as the name suggests, contains the plant's essence. It is this essence that makes herbs so powerful and helpful in so many ways. Fresh herbs used straight from the garden are highly aromatic and easy to grow. Alternatively, when the growing season ends, dried herbs can be stored for long periods of time.

Additionally, there are two traditional ways to dispense an herb's beneficial oils: through liniments or infusions. While the concept is similar in both methods (to steep or brew the herb), the carrier differs.

Liniments use rubbing alcohol to "carry" the essential oil. In regard to chicken keeping and coop hygiene, liniments may be placed in a spray bottle and used to fumigate the coop and roosts, reaching the housing's nooks and crannies. Liniments are not for topical or internal use, however, and should never be ingested. As a safeguard, always label your liniment containers "For External Use Only."

Infusions, on the other hand, may be used topically and internally for poultry. A neutral carrier oil, such as olive oil or coconut oil, combined with essential oils or herbs and gently warmed will serve a variety of purposes.

The following recipe is an easy infusion to make at home and a great way to introduce garlic into an adult flock's diet.

Garlic Oil Infusion   Serves up to 10 chickens

1 to 3 cloves of garlic (depending on your birds' tolerance), chopped or minced

¼ cup (79 ml) extra-virgin olive oil

2 cups (475 ml) layer feed

1 cup (240 ml) scratch

1. Steep garlic and olive oil in a glass jar for three to four hours, preferably on a sunny windowsill or in the sun.
2. Combine layer feed and scratch by mixing evenly in a "chicken-safe" bowl.
3. Drizzle the steeped oil infusion over the feed/scratch mix, and turn to coat.
4. Provide to birds immediately and use any leftover oil within 36 hours.

Raw garlic is powerful indeed, and you could say it's an acquired taste. I recommend getting young poultry off to the right start by developing a taste (or tolerance) for the flavor of garlic from an early age. Offer small amounts crushed in a separate dish with feed sprinkled over it or mixed into a mash (created by mixing some feed with water, yogurt, or applesauce). My chicks usually gobble it up when mixed with unsweetened applesauce. If you're tending to an adult flock and they haven't developed a taste for garlic, you can crush a clove or two and place it in the bird's drinking water. Remember to remove the cloves and replace dirty water often.

Honey. Without question, honey absolutely falls into the superfood category. In cultures across the world, honey has long been used as a treatment for wounds and burns, as a food preserver, and as a sweetener. Whole books have been written about the wonders of honey, and new research continues to indisputably prove its health benefits. When it comes to chicken care, honey works well as an anti-microbial "ointment" of sorts. It is perfect for wounds or burns, and it inhibits the growth of bacteria and may be combined with herbs such as yarrow and calendula to help speed healing.

Apple Cider Vinegar. Traditionally used as a folk remedy, the purported health benefits of apple cider vinegar include the cure to a variety of human ailments from allergies and acne to diabetes and high cholesterol. (In my first trimester of pregnancy, apple cider vinegar in a little water was the only thing that would alleviate my nausea.) Made through the fermentation of apple cider, where natural sugars are broken down first into alcohol and then into vinegar, apple cider vinegar contains several acids found in traditional vinegars, including the main ingredient, acetic acid. Take one whiff and you'll know it's powerful stuff.

Discourage bacterial and fungal growth in your chicken's water fonts by adding it to drinking water available for all poultry. Like most natural remedies, apple cider vinegar works on a preventive basis and doesn't take the place of good hygiene. Always scrub and clean water fonts thoroughly and replace dirty water frequently.

Apple cider vinegar works best as a digestive aid for young, growing birds. It assists in breaking down and digesting protein for better utilization. Particularly for growing chicks, balanced internal bacteria mean a healthier digestive system, which equals a bird that is able to maintain her health and fight off disease naturally. For best results, add approximately 1 Tbsp. (15 ml) of apple cider vinegar per 1 gal. (3.8 L) of water into a drinking container made of non-reactive material, such as plastic.

There are so many more beneficial plants and herbs with the potential to assist in chemical-free chicken keeping than are on this list (some are even considered weeds). Consult your veterinarian and a qualified herbalist for more about harnessing the impressive work of herbs in your chicken coop.

Dessert: Treats

A flock that is regularly offered treats will quickly learn to trust and follow their keeper(s); this comes in handy when training, breeding, preparing a bird for show, or administering medical care. Plus, giving treats is just fun—it gives you a chance to watch your birds, study their behavior, and get to know their individual personalities while they chirp, peep, and squabble over snacks.

Scratching the Surface

Scratch is the quintessential chicken treat. Each manufacturer of scratch will have a slightly different recipe, but the main ingredients are usually whole grains, such as cracked corn, wheat, and barley, with secondary ingredients such as sunflower seeds, dried mealworms, flaxseeds, and even dried fruit like cranberries and raisins. As you can imagine, scratch is like candy for chickens, and, in fact, a wise chicken keeper knows to consider it as such. Too much in the way of scratch and similar treats may lead to overweight hens and a host of health maladies. (Read "Chicken Obesity: Fact or Fiction?" on page 120 for more on the dangers of fat chickens.)

Scratch is best offered as a treat in the fall and winter when a flock is gearing up for the cold. A bit of extra body fat will help your birds combat the dropping temperatures and retain body heat. For the same reasons, scratch should be offered in moderation in the warmest months of summer. During that time, I reserve scratch as a special treat that works *for* me. If I need to examine an injured bird or lead the flock to a certain area or back into the coop, for instance, scratch will always keep them occupied.

Kitchen Treats

Just about any unprocessed food scrap from the kitchen can be given to your chickens. This includes scraps or cuttings from nearly every fruit or vegetable, with a few exceptions. (See "What NOT to Feed Your Chickens" opposite for those exceptions.) My flock adores fruit such as overripe bananas, peaches, apples, raisins, and every berry variety I've ever offered to them. Applesauce is a huge hit. Most chickens also love leafy greens, summer squash, winter squash, tomatoes, and other soft vegetables, too. Chickens also adore meat, and it's a great source of extra protein for certain times of the year, such as when the flock is molting. Just make sure bones are large enough to prevent swallowing and remove them from the coop after a day or so. Dairy products are also fine in moderation. In fact, yogurt is a great source of probiotics; when mixed with the crumbs and powder at the bottom of the feed bag, it makes a great treat while keeping waste to a minimum. Pasta, stale bread, and rice are tried-and-true favorites, too, as long as they are well cooked and offered in moderation. Cooked eggs and crushed eggshells are also well loved by chickens. As long as the eggs are cooked and the shells are crushed, this treat won't encourage egg-eating behavior.

Hand-feeding treats to your hens will build mutual trust.

Whatever the food item, just be sure that the pieces are a manageable size for a chicken to gobble up. Leafy greens, in particular, should be chopped or torn into smaller pieces. Chickens tend to get greedy and excited when they see the scrap bucket—a potentially lethal combination if an individual swallows large, whole chunks of food in a hurry. As long as the flock has consistent access to grit, they'll likely fare well without

What NOT to Feed Your Chickens: Common Toxic Plants and Foods

There are many plants that could pose a potential risk to your birds if ingested, but that doesn't necessarily mean they will. Some chickens will instinctively keep a distance from toxic plants (they tend to taste pretty bad, actually), and some will not. Plus, not every toxic plant is equally toxic. Some plants and their parts (which include stems, roots, leaves, flowers, and fruits) may pose health risks in varying quantities. A plant may have toxic qualities, but a bird would likely need to consume the entire plant for it to be fatal (and she would probably only do this if she were very hungry or confined to a space with no other options, such as indoors with houseplants). The best course of action is to avoid offering any plants that are known to be toxic to birds as treats and to find comparable alternatives to those toxic plants to sow in your garden or around the coop.

Below is a basic list of common flora that may be toxic to birds. There are so many potentially toxic plants they could fill their own book, so this isn't a comprehensive list by any measure. If you're unsure of a plant's toxicity, ask the horticulture experts at your local nursery where you purchase the plants, call your county extension office, or consult your avian veterinarian.

- Avocado (the pit contains a toxic fatty acid called persin which is fatal to all birds)
- Black locust
- Black nightshade
- Castor bean
- Corn cockle
- Jimsonweed
- Milkweed
- Monkshood, wolfbane
- Morning glory seeds
- Nightshade plant leaves (eggplants, peppers, potatoes, and tomatoes)
- Oleander
- Poison hemlock
- Potatoes, green or sprouted
- Rattlebox
- Rhubarb leaves (the oxalic acid contained in them is also poisonous to humans)
- Soybeans, raw (enzymes in the bean's raw form can cause digestive upset)
- Tobacco
- Yew

any undue digestive upset, but it's always better to be conscientious of the size of the treat you give them, just to be safe.

Above all else, use common sense when feeding kitchen scraps and other treats to your birds. Food should never be moldy, and everything should be offered in moderation. Processed foods, takeout leftovers, and other greasy or overly salty foods should be avoided. Send these to the compost heap or worm bin instead. Finally, make sure your birds have had their fill of their feed ration *before* offering treats. Not only will treats fill them up quickly and not meet all of their nutritional needs, but withholding their daily ration could also lead to hungry, eager birds that eat too fast.

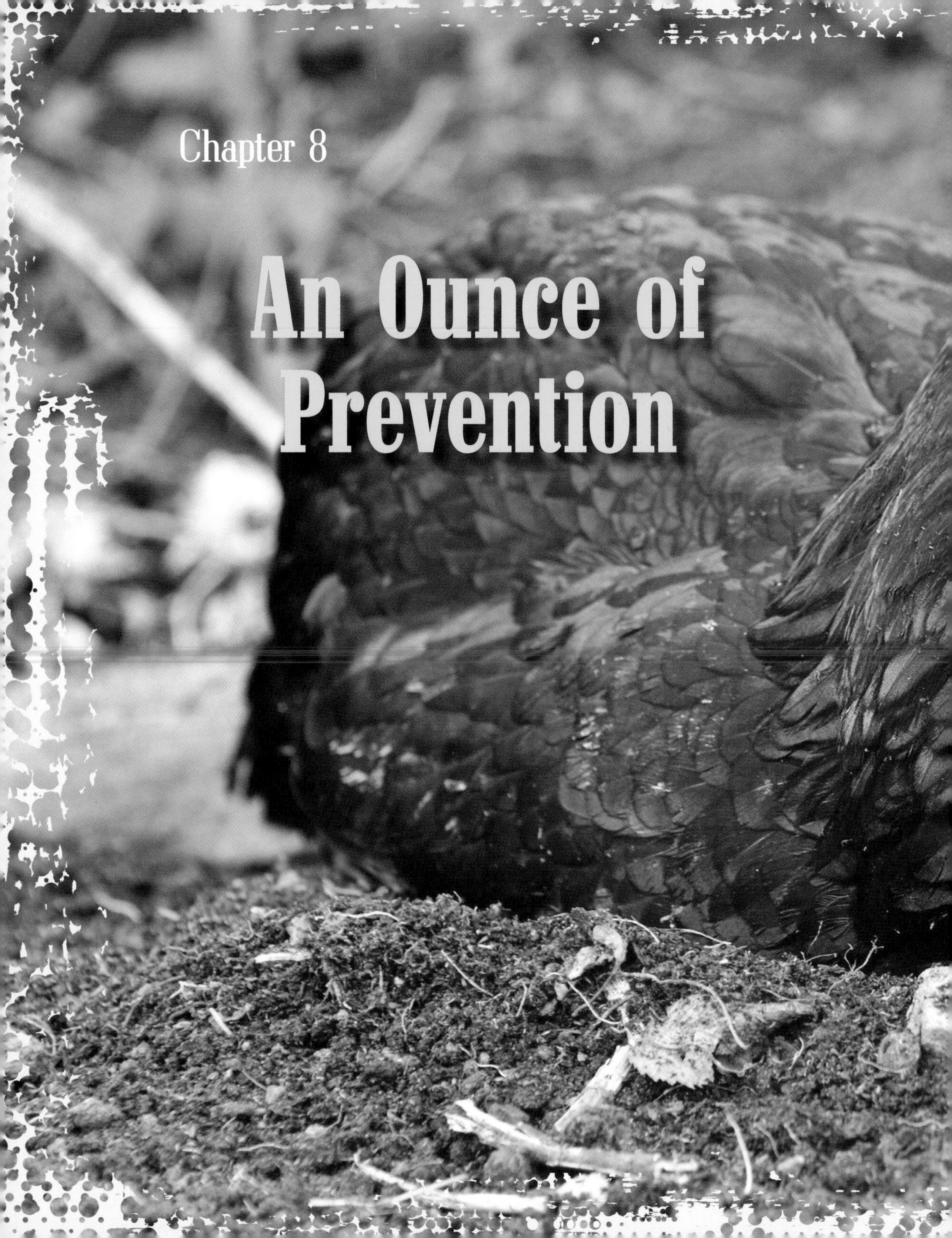

Chapter 8

An Ounce of Prevention

To become a chicken keeper is to become a bit of a detective. As animals of prey whose survival is tied to a strong social lifestyle, chickens don't like anyone to know when they're sick. In a flock, if a hen lags behind or looks weak, she makes an easy target for predators. If she displays signs of illness or lameness and the flock takes notice, they may well cannibalize or inflict further damage as a way to eliminate her from the group. While this may sound harsh, this is how chickens have survived when absolutely everything wants to eat them, from the tiniest mite to those with sharp teeth and talons. And since these traits will be apparent in even the tiniest backyard flock, it helps to keep a keen eye and a sharp sense as their keeper.

This chapter, as the title suggests, is all about taking preventive measures to keep your flock healthy. Chickens are incredibly easy animals to care for if you get them set up with all they need. As you'll read in a moment, I wholeheartedly believe that awareness and prevention are the two most reliable and inexpensive "medicines" of all, and that with these two tools, nearly every chicken ailment can be avoided. In my opinion, a healthy flock starts with mindful animal husbandry, and good animal husbandry starts with prevention. Preventive measures of good nutrition, conscientious chicken keeping, and a keen eye can often stave off some of the worst a flock may endure.

Chicken Health and Wellness

The chicken keeper's first tool—awareness—is largely about observation. What do you see when you watch your chickens? How do they act? How do they interact with each other? At first glance, your family of birds may appear to be ignoring each other, simply pecking around, eating, drinking, laying eggs, and grooming. If you look closely, you may notice subtle (or not so subtle) gestures between two or more birds as they communicate through vocal sounds, body language, and pecks, giving you a sense of hierarchy. If you watch even more closely, you'll get to know their individual personalities, where they fit on the hierarchical totem pole, and what their "normal" behavior looks like. How do they act when all is well? In order to be aware of any changes in your flock, you'll first need this frame of reference. Let's look at what both a healthy and sick chicken generally look like.

The Healthy Chicken

A healthy bird is aware and alert. Her body posture is erect, she has a robust appearance, and she carries herself confidently. She moves easily and freely and goes about her normal chicken behaviors—scratching and digging in the dirt, pecking and picking at what she finds, laying as often as her age and breed dictate, roosting at night, regularly dust-bathing, and making noises based on her activities, such as clucking and cooing while foraging or singing her "hen song" after laying an egg. She can easily move her body, flap her wings, and fly up to a perch or roost. Her behavior should be typical of her personality, submissive if she is lower in the flock's hierarchy or more dominant if she is an alpha.

For appearance's sake, a healthy hen is rather beautiful. Her feathers are shiny, her eyes are bright and clear, and her comb and wattle are a rich, deep red (unless her breed sports a different color). Her feathers lie flat against her body (unless she's a frizzled variety), her shanks are yellow or slate (depending on her breed, of course), and she is fully feathered (unless she is molting, which I discuss later in this chapter).

A healthy bird eats and drinks regularly throughout the day, her weight remains steady, and her droppings are fairly uniform. Droppings are a pretty reliable barometer of chicken health, so note any changes in appearance or odor. Firm, light brown to grayish green droppings with a white crest (the urine salts) are normal, with an occasional "cecal" dropping that resembles molasses in texture and color.

Chicken Lore

Roosters have many fine attributes. They protect their hens, show them where the best food is located, and keep eggs fertilized to ensure future generations. It's a popular misconception that a rooster must be present in order for hens to lay eggs. Thankfully for urban and suburban chicken keepers, nothing could be further from the truth. Hens will lay whether or not a rooster is present. However, if you want to hatch chicks from those eggs, you will need a rooster present to provide their mating services and to fertilize those eggs.

Pruning and preening are social activities that often take place in small groups within the flock.

The Sick Chicken

Identifying a sick chicken is a bit harder than identifying a healthy one. For starters, a sick chicken may still appear healthy on the outside and to the naked eye. That is why it is so important to regularly monitor your birds and know their typical personalities. When something is amiss, you'll likely spot it based on behavior rather than on appearances.

With that said, a sick chicken may still appear unkempt. Her body posture may lack strength, and she may sit idly for long periods of time. (Do not confuse this with broodiness; a broody chicken will be alert, making healthy vocalizations and likely defensive of her nest.) A sick bird's eyes may appear glazed and unfocused, and her comb, wattle, legs, and feathers may appear dull. She may not dust-bathe or preen and groom herself. A sick chicken usually stops laying eggs or lays eggs with severe deformities. She may stand very still at times, refuse to eat or drink, or lack the strength to get to the feeder or water font altogether. A sick chicken may puff up her feathers, may have clear leg or wing deformities, or may have obvious external pests. Droppings may change in color or consistency.

A healthy chicken will be perky and active and sport shiny feathers; bright, clear eyes; and a dark red comb and wattle.

A sick or injured chicken may drag a limb or bleed. If her injury is very apparent to *you*, rest assured it's very apparent to her flock mates. An obviously injured or ill chicken will likely fall prey to flock cannibalism. If you see excessive bullying, take this as a sign to investigate the health of the bullied chicken right away.

Finally, a sick bird will just not be herself. If you keep a small flock of laying hens and take the time to know each one's unique personality, quirks, and mannerisms, you can easily use this knowledge as a reference point on health.

So how do we stay ahead of disease, illness, and injury? Let's take a look at some preventive measures.

The Basics of Prevention

The second tool in the chicken keeper's most precious arsenal is prevention. How you tend to your flock, set up their housing, and care for them day to day will largely determine how well you can manage the spread of unwanted pests and illness in the long run. The following pages describe the measures you can take to give your flock the best chance at health and wellness.

Biosecurity

Biosecurity measures are the ones you take as keeper to prevent the spread of pests and disease in your flock. Here are some ways to practice good biosecurity:

- Exercise caution when adding new birds to a flock and use the quarantine methods outlined in chapter 11. The same goes for "chicken sitting" any outside birds.
- Have a designated pair of muck boots or Wellies for your coop and/or run and wear them only for chicken chores. Have a pair of "chicken boots" for each member of your family, and create a rule that children are only to wear them—and must wear them—while working with or caring for the flock.
- Wash hands and change clothes between handling birds from separate flocks.

You'll see more specific biosecurity measures listed under specific ailments in chapter 9.

One way to prevent the spread of pests and diseases is to have a designated pair of muck boots for each member of the family to wear only when caring for your flock.

Shelter

Adequate protection from predators and the elements is another basic tenet of keeping a healthy flock. Chickens aren't doing very well if they're regularly being eaten by foxes. Your flock's housing should be a well-ventilated shelter that is relatively clean (realistically, though, chickens make a mess of any structure they live in, so take the term "clean" here with a grain of salt). Fresh, dry bedding is a must, and waste should be removed periodically. Also, daily egg collection will avoid all manner of potential pests, predators, and cannibalism, such as egg eating.

Food, Water, and Supplements

Your flock should have access to fresh water, all day, every day, year-round. Water keeps your birds well hydrated, of course, but it keeps them laying, too. An egg is about 74 percent water, meaning it requires quite a bit of water just to produce it. Without water, not only will egg production suffer but a flock's collective health will also decline rapidly. Your flock should also have access to age-appropriate food at all times. Supplements, such as grit and oyster shell, are critical for good digestion and a functioning reproductive system (respectively). Just having those two supplements available to your flock will stave off most kinds of digestive and reproductive issues. (Look for more details on water, feed, and supplements in chapter 7.)

Stress Management

We could all do with a lot less stress, and poultry are no different. They respond to stress in much the same way we do, with a loss of appetite, lowered immune function, and increased fatigue, to name a few. How we tend to our flocks determines to a large extent how much stress they're under.

Whether from fear, overcrowding, the addition of new birds to the flock, or simply boredom, stress can manifest itself in unpleasant and unhealthy ways. Fear, for instance, is a relatively common

Vaccinations

As with vaccines designed for humans, dogs, and other pets, vaccines for chickens are inoculations designed to trigger immunity in the recipient against a given disease. The most popular vaccines available for poultry include Newcastle disease, coccidiosis (six out of the nine strains known to affect chickens), Marek's disease, fowl pox, and infectious bronchitis, among others. Vaccines can be administered by injection or through orifices, such as the eyes, nose, or mouth.

How do you know which diseases to vaccinate for? Or should you vaccinate your chickens at all? First and foremost, determine if a given disease is prevalent in your area. Seek out your local chicken club and ask members to share their experiences with communicable diseases and which (if any) they chose to vaccinate against. In my experience, a backyard keepers club will have the most recent information on diseases in your area; however, you may choose to call your county extension office to get a broader history of poultry diseases in your greater region. Then, research the individual diseases and their vaccines. Chickens should only be inoculated against diseases they have a reasonable chance of contracting.

If you choose to vaccinate, when is the best time? Depending on the disease and the vaccine itself, some vaccines must be given within the first 24 hours of life. Others may be given at various life stages. If you choose to purchase your birds from a hatchery, they will likely give you the option of vaccinating your day-old chicks before having them shipped to you. The cost per vaccine is rather reasonable, about 15 cents per dose, give or take, depending on the vaccine. Some vaccines require only one dose; others require boosters. Some backyard flock owners or breeders may choose to inoculate their own birds. If you take this route, it's recommended that you do so under the supervision of a trained professional—your avian veterinarian. Furthermore, vaccines are sold in very large quantities (500 to 1,000 doses), which is usually not realistic or economical for the backyard flock owner. Instead, if vaccinating is important to you, consider starting your flock from a source that reliably vaccinates, such as a hatchery or reputable breeder.

experience for chickens. As creatures of prey, they must constantly be on the alert for danger. Whether the fear is from a legitimate source, such as a predator, or from an imagined threat, such as a loud noise, the stress response is the same.

Severe overcrowding or the addition of new birds to an established flock can lead to picking, pecking, and cannibalism of varying degrees, from minor bullying to fatal wounds. Boredom itself is rare in chickens; if a flock is confined to a small space without room to do what they naturally do, "boredom" ends up looking a lot like cannibalism.

So, reducing your flock's stress first means knowing their nature (as we just learned) and then adjusting your activities accordingly. Chickens are not picky. They don't ask for much. If you simply allow them to tap into their natural behavior and express these traits, they will be adequately entertained and quite content. Here are a few ways to do that:

- Many issues of cannibalism in an otherwise healthy flock can be warded off by providing space to roam and giving the freedom to *be* a chicken. Of course, this means having enough physical room in the coop or enclosure. But this also means having a place to dust-bathe and scratch around, doing the things that chickens do. With enough space, chickens feel less territorial; submissive birds have the ability to move away from any roosters or alpha hens that want to assert their dominance. In other words, with the ability to focus attention on normal chicken behaviors (like grooming and foraging), less attention will be focused on flock mates.

- To reduce the stress of integrating new birds into one flock, follow the recommendations suggested in chapter 11. There are many ways to go about it and while not a completely stress-free endeavor, most birds do get along quite well in the end.
- The stress caused by imminent predators is hard to control in most aspects, but you do your part by providing a safe coop and making sure the birds are securely locked up each evening. Take the precautions against predators explained in chapter 6, such as burying perimeter fencing or using electric fencing. If your town or city allows it, consider adding a rooster to your flock; aside from fertilization, a rooster's main occupation is security guard. A good rooster will be willing to risk his life protecting his flock, making the hens feel more secure day to day.
- Finally, minimize noises in and around your birds' housing when they're inside. Schedule mowing, weed whacking, leaf blowing, and other outdoor maintenance when your flock is free ranging or otherwise occupied. If they can comfortably move far away from the noise, they will be less frightened. Especially try to avoid such chores early in the morning since the stress of loud noises may impact a hen's egg laying.

Chicken Lore

Old wives' tales concerning roosters are abundant. A common misconception is that roosters crow only at daybreak. While they ***do*** crow at dawn, they also crow at morning, midday, afternoon, evening, or anytime they hear an alarming sound, think a predator may be near, want to assert their dominance. . . you get the idea.

Dust-Bathing

The health of a chicken's feathers can greatly influence her overall well-being and her ability to ward off external pests. As a keeper of chickens, it's your job to give your flock the opportunity and space they need to maintain pristine plumage.

Chickens maintain their feathers' good health by taking dust baths, an entertaining series of romps in the dirt (for chicken and human alike). After digging a shallow hole, your bird will fluff herself sideways, tossing dirt and sand deep down to the base of her feathers. Once there, the dirt will asphyxiate pests that may prey on her. After bathing, the bird preens. She sources oil from the *uropygial* gland (also called the preen gland) located at the base of her tail and distributes this oil among her feathers with her beak. This oil creates sheen, promotes health, and ensures water resistance in the feathers. Birds that are debeaked are unable to properly gather oil and preen, yet another reason *not* to buy debeaked chickens or to debeak them—ever.

This hen is enjoying a dust bath, one way that chickens stay calm and keep their feathers healthy.

Though a flock may often dust-bathe and preen in pairs or all together, each individual chicken will self-regulate her grooming patterns and hygiene, repeating the dust bath and preening only as often as she finds necessary.

Dust-bathing and preening are essential to good health, and as the flock's keeper, you can provide optimum space and material for them to bathe in. If your birds lack an area in which to dust-bathe (or if you're tired of their using your potted petunias), create an area yourself using the instructions below.

What You'll Need

Builder's sand (available at your local hardware store)
Wood fireplace ash
100 percent food-grade diatomaceous earth (DE)
1 mini–raised bed, galvanized steel basin, or low plastic bin: minimum size 18 in. x 24 in. x 10 in. (45 cm x 60 cm x 25 cm) deep or more

1. Combine equal parts sand, wood fireplace ash, and DE and mix thoroughly.
2. Pour mix into your chosen container, leaving about 4 in. (10 cm) of headspace between mix and rim of container (to catch excess as it is tossed).
3. Replace mix as needed, so there is always at least 6 in. (15 cm) of material in container.
4. Offer this dust-bathing area to your flock all day, every day, all year long.

These three chickens are happily dust-bathing after digging shallow holes in the sand. Having adequate space to dust-bathe is essential to good health.

Molting

When I first began raising chickens, a farming friend and poultry guru warned me to keep an eye out for my flock's first molt, probably in an attempt to spare me the inevitable panic when I saw my birds in such a state. Unkempt, disheveled, and bedraggled birds are what I found that first November, and even with a seasoned poultry keeper's warning, the sight was shocking. Luckily, there was no need for me to worry, and you shouldn't either. Molting may look off-putting at first glance, but it's a natural, normal, and painless part of life with chickens.

What It Is. Molting is the annual process of shedding and regrowing feathers that all birds go through. Molting usually happens pretty predictably in autumn, so there's plenty of time to prepare.

Your chickens will typically molt during their second fall. The shorter day length signals the bird's system to prepare for the coming cooler weather and to replenish plumage. Molting can begin anywhere from late summer to early winter and may take between 4 to 12 weeks to complete, depending on breed, age, and the health and vigor of your individual birds. Each one is different.

Though molting is triggered by waning daylight hours, it can also be triggered by stressful situations, such as going without food or water for a period of time. To safeguard against unnecessarily prompting a molt, make sure your flock has the "basics" covered. Expect your birds to molt annually. Your best layers and healthiest birds tend to molt quickly and efficiently, returning to laying within a few weeks.

It's important to remember that molting is normal, so don't be alarmed when you see the feathers fly. If birds have adequate space in their coop, run, or pasture, molting will not induce pecking or other cannibalistic behaviors. If your flock experiences extreme feather

Wing Trimming

Adult chickens are pretty pathetic flyers in general, but when motivated (such as when being chased by a predator or when eyeballing a particularly tasty treat just over a fence and out of their reach), they can put their wings to use. Wing trimming is the practice of cutting back a chicken's wing feathers to reduce her ability to fly. The clipped wings only hinder the flight of a chicken until she goes through a molt and regrows new feathers. When done properly, wing trimming is not painful, but there is controversy on whether it is a necessary procedure. Many backyard chicken keepers prefer to erect proper fencing to keep chickens contained rather than manipulate their wing feathers. This is the route that I most strongly recommend. Personally, I feel that a bird's ability to evade predators in an emergency situation (and for her to comfortably fly up to her roost each night) is more important than keeping her out of a garden. Plus, it is the keeper's responsibility to contain the flock in a safe enclosure and protect his or her personal gardens from chickens' eager beaks.

picking and pecking (evidenced by drawing blood), however, they probably need more space. Other than a few small behavioral differences (more on those to come), your flock should act normally during the molt.

What It Looks Like. Just before molting, a bird's plumage becomes dull, losing its sheen as the old feathers lose health. Shedding begins at the head and works gradually back toward the tail. Each bird has her own molting "style"—some lose all plumage at once; others lose feathers so gradually and subtly, you'll wonder if they're molting at all. As the process unfolds, so to speak, you will see small pinfeathers emerging from areas around the chicken's body as new plumage grows; these appear as tiny quills at first and eventually open to reveal the fresh, new feathers.

The easiest way to spot if a molt is underway is to take a look around the coop. If you find feathers everywhere and have ruled out a predator attack, you can probably safely assume your flock is molting.

What You Can Do about It. Molting is really hard work. Regrowing feathers requires a lot of energy, so expect to see a dramatic drop in egg production for a while. The hen's body pulls energy away from the reproductive system and puts it toward the growth of new feathers (as well she should; feathers are approximately 85 percent protein).

While this can be disappointing for the keeper of a laying flock, keep in mind that your bird's overall health is critical to all future egg production, and feather health is of the utmost importance. With healthy feathers, adequate dust-bathing and preening opportunities, your bird will be able to keep mites, lice, and other external pests at bay.

Forced Molting

Not surprisingly, large commercial egg operations function under the mantra that higher egg numbers equal higher profits and will subsequently "force" a molt onto their birds to trigger one last boost of egg laying before taking those birds out of production indefinitely. A forced, or controlled, molt puts significant physiological and environmental stress on the birds. Most variations of forced molting include food and water deprivation for several weeks, the feeding of nutritionally deficient rations in minute quantities (such as oats or scratch), and significantly reduced light exposure (limited to eight hours a day or less). This environmental and nutritional stress triggers a short and fast molt—as little as three weeks—after which time the birds resume laying supposedly higher-quality, larger eggs.

There are small poultry flock versions of a forced molt, none of which I recommend. In short, a forced molt leads to a host of other health hazards, such as reduced immune function and increased salmonella susceptibility—not to mention, it's cruel and inhumane. If egg production is of the utmost importance to you, first research and then invest in the best laying breeds for your personal flock. When your birds do inevitably molt, there are a few ways you can support the process without losing too much production time.

So how can you help?

Consider offering supplements that contain a high percentage of protein. Mealworms are a fun treat and help to build trust with your birds. Kids will love to hand-feed treats. Better yet, allow your flock extra free-range time to find their own buggy source of protein. Some flock owners switch to a commercial broiler (meat bird) feed during a molt. Others give canned cat food, canned tuna, cooked ground beef, or other meat-based protein to their flocks during a molt as a supplement. This practice is fine in moderation but don't offer so much that your chickens turn up their beaks at regular feed. The balanced commercial layer ration should still be the primary source of nutrients in their diet.

These scruffy-looking hens are going through their annual molting process, the natural shedding of old feathers and the growth of new ones.

Further support your birds' molt by reducing their level of stress. In addition to taking the preventive precautions discussed earlier in this chapter, consider rescheduling shows or other public events with your molting birds—chances are you won't win any ribbons with your birds looking the way they do anyway.

A Word about Debeaking

The short version is: Don't do it. Debeaking is a practice used by large commercial egg operations (you may know them by the name factory farm) to keep their birds, crammed six or more to a small cage, from cannibalizing each other. The beak is incredibly important to every aspect of a chicken's life. Socially, birds use their beaks as a form of communication, through picks and pecks. Nutritionally, chickens use their beaks to gather feed, forage for bugs, and tear up large pieces of food. Hygienically, birds use their beaks to gather feather oil from the preen gland and groom themselves. Apart from all of these reasons, debeaking is painful, cruel, and inhumane. Refuse to purchase birds from breeders or hatcheries that routinely debeak, and explain why you won't be purchasing their birds or supporting their management practices.

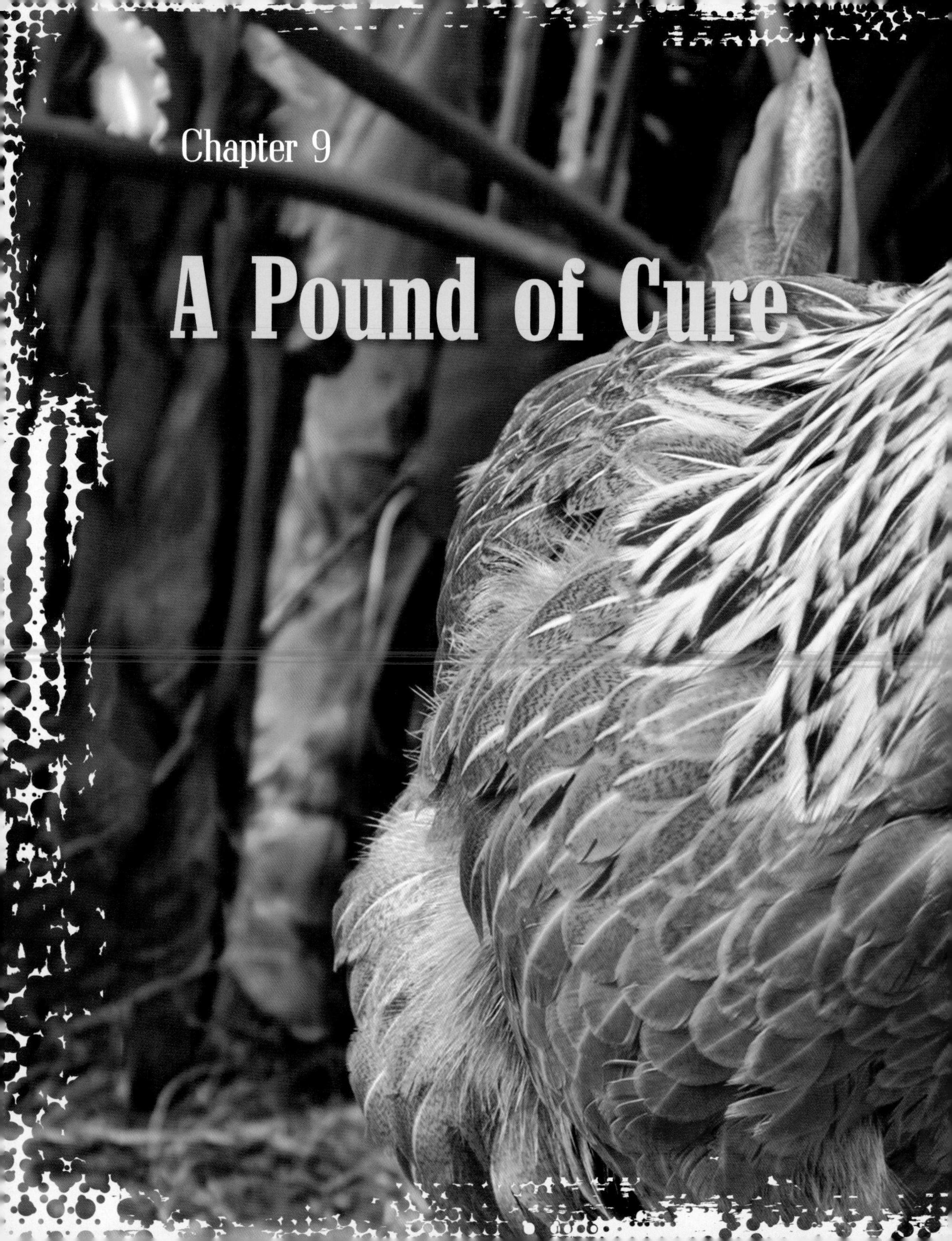

Chapter 9

A Pound of Cure

Now that you know what it takes to keep a flock healthy, I can assure you the alternative (treating birds when they get sick) is much harder. But, if we're being honest with ourselves, we won't always fend off every mite, every broken toenail, every bully. To some degree, it's inevitable that we will have to care for the hurt, sick, and dying members of our flock. That's just part of life. As such, this chapter focuses on the more prevalent pests, diseases, and injuries you can expect to encounter in your avian adventures. There are many others out there, but most are very rare and some manifest only in very large chicken operations where conditions are cramped and unsanitary...and that won't be you or your flock.

On the following pages, these common ailments and vexes are fleshed out with definitions and descriptions of symptoms along with suggestions of what you can do about them. Some remedies are natural; some are conventional. With that said, please note that I am not a veterinarian; the recommendations here should be practiced under the supervision of an avian health care professional and, as always, exercise common sense and good judgment. Outcomes described are largely anecdotal (meaning I learned by making mistakes along the way). Always consult your vet if you have questions, concerns, or need treatment for your pet chickens.

Sick as a Chicken

Despite their exceptional hardiness, chickens are the kinds of animals that get sick rather quickly and go downhill fast, not always giving the keeper much time to intervene. Being both prey animals and flock animals, as they are, they tend to hide illnesses until the very end. An active observer with a keen eye will be able to notice the signs before it is too late (this is why it is so important to know your birds). Even without a keen eye, some problems are so obvious you can quickly jump right in to help your bird.

What You'll Need

Generally speaking, there are a few things to do as soon as you think you have a health and wellness issue on your hands. First, it's wise to immediately isolate any bird you suspect to be ill or injured. Isolation will help in several ways. First, it allows the bird to relax and not stress about the flock bullying her; second, it effectively keeps the flock from bullying her; and, third, it allows you to monitor the important things, like her food and water intake (or lack thereof), her droppings, and her ability to move on her own. Have an extra mini-coop, tractor, or pen to use as a sick bay for this reason. The ideal temporary housing gives you easy access to the sick chicken, allows you to assess the situation, monitor the bird's progress and treat, if necessary. I have a movable tractor that I've used to raise chicks, let a broody hen sit in peace, or employed as a sick bay. But, if I've had to bring a bird inside the house for comfort, a cat or dog crate has worked well. I'm not ashamed to say we've even used a cardboard box in a pinch. If it covers the basic needs of housing (comfortable, dry, well-ventilated, and safe), then it fits the bill.

Finally, you'll want to have a chicken first aid kit prepared and at the ready should something happen. You can add many more items based on your region and flock, but here are the basics:

- Antiseptic cream
- Chick water font and feeder (enough for food and water for one chicken)
- Hemorrhoid cream
- Hydrogen peroxide
- Mini-coop, tractor, pen, or pet crate with a nest box for hens
- Nail quick stopper
- Nail trimmers
- Non-petroleum jelly or vegetable oil
- Rubber gloves
- Scissors
- Scratch and/or mealworms (a tasty treat for distraction)
- Sterile gauze and surgical tape
- Washcloths (for chicken-use only)

If you suspect there's something wrong with one of your hens, it's wise to isolate her. You can create a simple portable sick bay using a movable tractor, pet crate, or even a cardboard box so you can monitor the bird's progress and treat, if necessary.

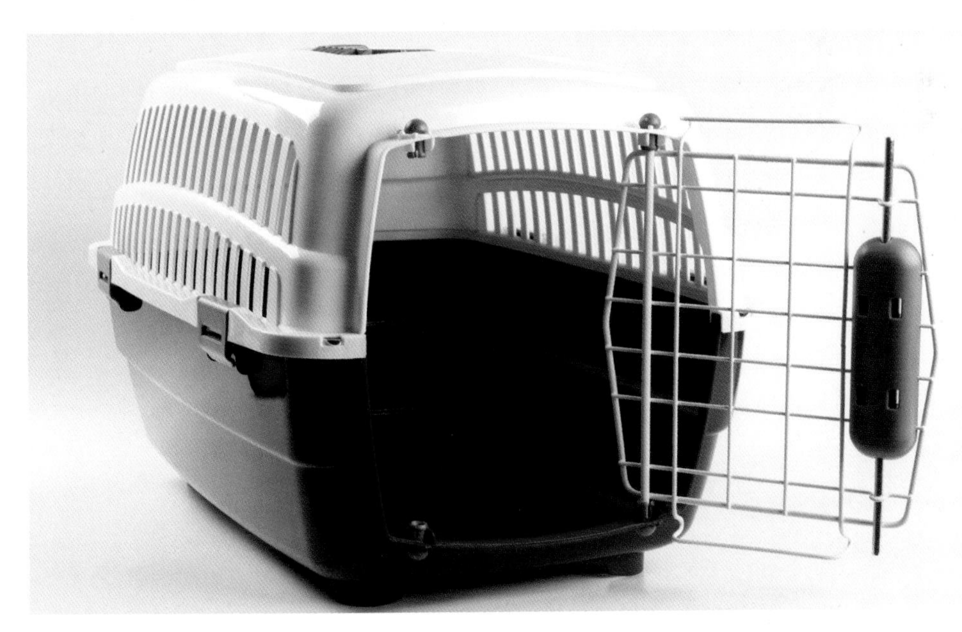

Tales from the Coop

The summer between my two years of graduate school, I traveled to Moshi, a small town at the base of Mt. Kilimanjaro in Tanzania, for an internship. I worked in a children's home with kids and teens of all ages and also in a small nursery school, packed to the gills with eager little ones ready to learn. We would often take breaks with the younger children and walk to the market in town to pick up food and supplies.

During one afternoon at the market, I noticed women buying and selling chickens. When the exchange was complete, the buyer scooped up her hens by the feet and carried them away upside down, several in each hand. I was astonished at how still they became once they were hanging upside down: Squawking, flapping, and general discontentedness seemed to disappear. For a second, I even thought they might have been dead.

Years later, when I had my own flock, I confirmed that carrying a chicken by her feet would indeed quiet the bird. It worked, and it was easy. But it's not a carrying method I recommend. I learned that when a chicken is inverted, her internal organs put pressure on her lungs, making it difficult for her to breathe. More than likely, your birds are also pets, or your flock belongs to the family. Building trust with your hens, then, is probably quite important to you. For those reasons, the best way to carry one of your birds is the method described in this chapter.

Pests

Chickens can help keep your landscape free of slugs, bugs, and weeds, but they are susceptible to their own parasites as well.

What They Are

For their tiny size, poultry lice and mites can make life pretty uncomfortable for your birds.

Lice

This parasitic insect lives on and chews the skin of chickens. To be more precise, there are two main categories of louse: *bloodsucking*, which targets only mammals, and *chewing*, which targets mammals and birds. Of the chewing kind, there are several varieties that attack chickens, and they are species specific, meaning they'll only infest chickens. A lice infestation has some undesirable and very frustrating consequences (more below).

Mites

A relative of the tick but resembling a small spider, the mite is a tiny (less than 1⁄25 of an inch) parasite that can wreak havoc on a chicken's skin, feathers, and immune system. There are several varieties of mites: Most live on the body of birds and survive by chewing skin or feathers, or sucking their blood. Some types may even burrow through the skin into the body of the bird. Some mites are easier to spot with the naked eye (and fast reflexes) than others. The most common mite in backyard flocks on the North American continent is the red mite. Also called poultry mites, roost mites, or chicken mites, these buggers make their home on the flock's roosts, coming out at night to suck blood while the birds sleep, changing from gray to red in the process. Populations can explode in warm climates or in the heat of summer.

What They Look Like (in Your Flock)

Lice

A lice infestation may result in lowered fertility and decreased egg production, feather damage, lack of restful sleep, and decreased appetite leading to malnourishment. A resulting weakened immune system is common, which can leave your flock susceptible to diseases and internal parasites. Lice and nits may be visible on your birds at the feather shaft, or you may simply notice your chickens shaking their heads or scratching their ears in irritation.

Mites

A mite infestation may result in increased appetite but lowered egg production, feather damage, and anemia. Especially weak birds, such as chicks, may die or fail to thrive from a heavy infestation. Chickens will often pluck out their own feathers in an attempt to stop the irritation of being bitten, so on infested birds you may see featherless areas on the bird's body, bare skin, bloody spots, or scabs. Red mites may be visible as tiny red specks on the chickens or on roosts at night.

What You Can Do about Them

Prevention

Proper coop management and cleanliness is paramount in preventing mites and lice from getting a foothold in your flock and spreading among your birds. Practice good biosecurity by...

- Keeping new birds in quarantine before integrating with your flock.
- Washing your hands when handling birds from other flocks.
- Changing clothes and shoes worn around other chickens. (Designate a "chicken-coop-only" pair of boots for chicken chores for each member of your family.)

Dust-bathing is a chicken's best defense against mites and lice.

Provide a dust-bathing area for your flock throughout the year. The ash from a fireplace or woodstove is a particularly helpful ingredient for homemade dust baths (see the recipe on page 136), as is diatomaceous earth (DE), your new pest-preventing best friend. Use DE generously in nest boxes and around your flock's housing. You can even directly dust your birds once a month or so. Just be sure to wear a face mask and eye protection and do it outdoors since the fine powder can cause irritation if you get it in your eyes or nose.

Treatment

There are poultry-approved insecticide powders that can be used to treat adult lice (not nits) over the course of several weeks and multiple applications (as the nits hatch), but as a natural chicken keeper and a beekeeper, I can't recommend them in good conscience. DE is a safe alternative, but its use will take several more applications over a longer period of time to be effective.

Mites, thankfully, are environmental, and so they're easier to manage or eradicate. Thoroughly clean the flock's housing, focusing on cracks and crevices and apply DE throughout. Treatment can be time-consuming, physically taxing, and stressful on your birds, but you can get rid of them.

Diseases and Illnesses

Chickens are susceptible to a large number of diseases and illnesses; most are mild, but some can be more serious. In this section I discuss the most common ailments your flock might encounter—their symptoms, ways to prevent, and treatment options.

Bumblefoot (Foot Pad Dermatitis)

What It Is

Also called foot pad dermatitis, bumblefoot is a common condition that occurs when an abscess on the foot pad becomes infected with the *Staphylococcus aureus* bacteria. Heavy breeds are most likely to suffer from bumblefoot, although chicks can just as easily get it under the "right" conditions. Those conditions include unsanitary or damp housing, splintered roosts or roosts with sharp edges, spending excessive amounts of time on wire-bottomed flooring, and any other conditions in which the bird may injure the foot and introduce bacteria into the wound.

Adequate free-range time can go a long way toward reducing the risk of bumblefoot and other ailments.

What It Looks Like

A chicken with bumblefoot will limp, show lameness, or have difficulty walking. If you look on the underside of the foot and notice the foot pad is red, swollen, or filled with pus or scabbing, the bird likely has bumblefoot. You might also see sores on other areas of the foot, such as on the bottoms of the toes. Left untreated, these symptoms will continue to worsen.

What You Can Do about It

Prevention. Good sanitation in the coop and enclosure, fresh, clean bedding, and adequate ventilation will go a long way in preventing the spread of bacteria that can lead to bumblefoot. Birds may injure their feet at any time while free ranging, jumping up and down from roosts, and just by being a chicken, but reducing the number of potential obstacles will seriously help to prevent this condition:

- Don't house your birds on wire-bottom flooring.
- Don't use gravel or any large, sharp rocks for bedding or the floor of coops and runs.
- Sand down wooden roosts until smooth.

- Use appropriate bedding materials in the coop, run, or other enclosures.
- Remove and treat any birds that have bumblefoot to reduce the spread of bacteria and eliminate the chances for bullying.

Treatment. Treatment for bumblefoot is rather unpleasant and not always effective. First, isolate the injured bird and put her on dry, clean, soft bedding. Some chicken experts recommend cleaning the area, cutting the abscess, and removing the fluids inside, followed by a hydrogen peroxide wash, antiseptic cream application, and sterile bandaging. If you choose this route, I'd recommend doing so under the supervision of a vet, who may show you how to do it the first time and be available for questions as treatment continues. With this method, bandages will need to be removed and replaced every two to three days.

Note of Caution: There is a minor human health risk when handling bumblefoot in chickens: impetigo. This superficial skin condition is very contagious, so practice excellent hygiene when treating your birds. Wear gloves, wash hands thoroughly afterward, and dispose of wound dressing by burying or burning.

Chicken Lore

If it starts to rain and chickens run for cover, chicken folklore says it won't rain for very long. If they continue about their business out in the open, the rain will linger all day. To be perfectly honest, this myth has a ring of truth to it when it comes to my flock.

Coccidiosis

What It Is

Coccidia are protozoan parasites that live in the ground, where they are picked up by birds (wild and domestic) and colonize and infect the intestines of your domestic chickens. Ground-dwelling fowl are naturally exposed to the parasite's eggs in their ranging, and a reasonable number of coccidia are usually present in the guts of most birds. The condition coccidiosis occurs when the parasites multiply, and the population colonizing the intestines gets out of hand, causing illness.

There are nine species of coccidia that can infect *Gallus gallus domesticus*, only six of which are covered in the current vaccination regimen administered to day-old chicks and offered by most commercial hatcheries. Younger birds are most prone to falling ill and dying from coccidiosis. Older birds may be infected and regularly shed coccidia eggs without showing symptoms, a condition called coccidiasis.

Other livestock (such as goats) and poultry may become infected with coccidia and fall ill, but thankfully, the strains that affect chickens do not affect other animals (including other poultry), and vice versa.

What It Looks Like

When a young chicken falls prey to coccidia, the intestine struggles to absorb nutrients, resulting in a malnourished bird. In an early case of coccidiosis, the chick may exhibit loose, watery stools; in an advanced case, blood may be present in the droppings. The presence of blood in the stools is an indicator that the case is quite serious, and many birds do not recover from that stage. Young chicks that do survive may fail to thrive; older birds that survive will often show a decrease in laying. For this reason, coccidiosis is often difficult to diagnose before it is too far progressed.

The worst cases of coccidiosis occur in young birds during their fourth to fifth week of life, although infection is most common between three and six weeks of age.

The most common environment in which coccidia flourish is damp, humid, warm and/or dirty, overcrowded, and unsanitary housing, where the bird is overwhelmed with the parasite. Because the parasites live everywhere in the ground, most chickens will develop gradual, lifelong immunity provided they mature in clean housing environments or on dry pasture.

What You Can Do about It

Prevention. A properly maintained free-ranging flock is less likely to become infected than birds living in crowded conditions. Monitor your hens frequently and follow these eight tips to keep your birds healthy:

1. Although it sounds unseemly, do monitor your bird's droppings with some frequency. Healthy droppings (relatively firm and formed green stool with a white crest) are typically good indicators of your bird's health.
2. Be diligent about keeping your flock's home clean. Watch the weather and be aware of when your coop, run, or pasture goes from clean and dry to dirty and soggy, especially when keeping younger birds with immature immune systems. Raise chicks that are hatched early in the year so they grow and develop immunity during cooler, springtime weather.
3. Work to keep water fonts clean and free of droppings. Your entire flock, if ground-raised, is exposed to a manageable number of coccidia; unsanitary conditions, especially feces in the water source, expose the birds to the overwhelming population of coccidia that leads to illness. The best way to keep birds from perching on water fonts is to hang the font from a chain at the bird's back height or purchase fonts made with an angled top to keep them from roosting.

Allowing your chickens to free-range on pasture will boost their immunity.

4. To help your birds acquire natural immunity, allow them to access dry pasture from an early age. Giving your birds ample time free ranging outdoors gives them a chance to develop gradual immunity as they are exposed to the coccidia parasite(s) that live in your area.
5. Vaccinations are available for day-old chicks, offered either for a small fee (15 to 20 cents) through a commercial hatchery upon purchase or in bulk packages (1,000+) that you can administer yourself. As with most poultry vaccinations, the vaccine for coccidiosis must occur within the first 24 hours of life. Keep in mind, however, that even vaccinated birds are still at risk for contracting and becoming overwhelmed by the parasite.

6. As an alternative to the vaccination, medicated feed containing a drug called coccidiostat is also available for young chicks. Coccidiostats may be in the form of medication added to feed or drinking water and must be offered during the first few weeks of life. While this medication may prevent a chick from contracting several strains of coccidia, it is not designed as a cure or treatment.
7. Whether you choose to medicate or vaccinate, or not, all chickens benefit from the addition of healthy probiotics to their daily diet. Offer cultured yogurt mixed with applesauce to young chicks (my elderly birds still love this treat). Also, many chicken keepers add apple cider vinegar to their flock's drinking water: 1 Tbsp. (15 ml) to 1 gal. (3.8 L) of water. Just remember, supplements should never take the place of an age-appropriate feed offered daily.
8. Proceed with caution when integrating adult birds into an existing flock. The strains of coccidia vary from place to place. While each bird is immune to the strains in her home environment, bringing adult birds together could introduce strains that either flock is not immune to, putting all in jeopardy.

Treatment. Although bloody stool often indicates an advanced case of coccidiosis, it's a rather reliable indicator that something is wrong. Take note of where your birds roost at night and acquire a stool sample from the bird or birds you suspect to be infected. Take this sample to your vet to confirm the coccidiosis diagnosis and identify the particular strain. He or she will guide you toward the appropriate treatment for that strain (treating the wrong strain could have serious consequences). Many medications to treat coccidiosis are water soluble and can be administered through drinking fonts, although this does require that you not eat the eggs for some time during and after treatment.

Sour Crop and Impacted Crop

What It Is

Sour crop occurs when the contents of the crop fail to empty fully, and the remaining food ferments, causing illness to the bird. As the chicken continues to eat, the crop bulges, yet no nutrients get through to the digestive system. Similarly, an impacted crop is caused by a blockage in the crop (such as a bird having ingested large blades of grass, feathers, or bedding) and may develop into sour crop. Though different, the two conditions have similar repercussions and many chicken keepers address them interchangeably.

If left untreated, crop complications can be fatal, but luckily, neither condition is contagious between flock members.

What It Looks Like

Some chicken maladies are as unpleasant as they sound, and sour crop is one of them.

A few telltale signs of sour crop include:

- Full crop that has failed to empty overnight. (A crop should empty every 24 hours or so.)
- Watery feeling to the crop (think of the consistency of a water balloon), or a crop that feels hard and solid.
- Foul smell emanating from the bird's beak (from rotting and fermenting food).
- Drop in appetite and little or no consumption of food.
- Jerky movement of the head as the bird tries to dislodge items in the crop.
- Natural isolation from the rest of the flock.

If you suspect sour crop, first pick up the bird, holding her securely with wings close to her body. Feel the crop (the area between the neck and the chest) for impaction. Next, gently open the bird's beak and take a sniff (you may need an extra set of hands to help). The scent of sour crop is not universal, since it depends on what the bird has recently eaten. However, it will smell distinctly unpleasant and...well, sour.

Remember that a full crop in the evening is normal. You may notice a bulge between the bird's neck and chest in the twilight hours—this is to be expected in a healthy chicken. If this bulge has not reduced in size by the next morning and the bird exhibits other symptoms as listed on page 149, an impacted or sour crop may be a possibility.

What You Can Do about It

Prevention. Sour crop prevention is easy, and when you're managing your birds properly, there is nothing extra or out of the ordinary to do. Continue the following three good management habits:

1. Provide grit. To avoid digestive-related issues, always provide a source of grit, even if your flock free ranges. If they need it, they'll take it; if they don't, they won't.
2. Feed live cultures. Yogurt is an excellent probiotic and when mixed into mash form with the leftover "dust" from their feed, it makes a great treat, too. Also, as noted earlier, add apple cider vinegar to the flock's drinking water; 1 Tbsp. (15 ml) per 1 gal. (3.8 L) should suffice as a preventive treatment. Do not add if you suspect a case of sour crop is already underway.
3. Withhold large treats or bones. To prevent crop blockages (or if you have birds prone to blockages), opt out of giving your birds treats with bones in them. Also exercise caution with large pieces of meat or produce that they're unable to break up with their beaks. This includes tough stalks or large, leafy pieces, such as whole kale leaves.

Treatment. As with most poultry maladies, isolate the affected hen as soon as possible. Monitor her droppings and water intake by offering water and/or vegetable broth in a familiar watering font. Do not offer feed at this time.

To actively loosen an impacted crop, you may opt to feed your bird a little vegetable oil—up to 5 ml per day—through a small dropper; oil may lubricate the crop's contents and aid in its movement. Organic olive or coconut oil works great. Again, using a second set of hands, insert the filled dropper into the back of the bird's mouth and slowly release the oil. Take great care to pass the small hole at the base of the bird's tongue—this airway leads to the lungs and if filled with the oil, it may cause the bird to choke. After administering the oil, gently massage the crop in a downward motion. Leave the bird in isolation with water and/or vegetable broth available overnight.

On the second day, repeat the oil treatment, as needed. Provide very soft foods, such as buttermilk-soaked bread in small pieces (the buttermilk may have some active cultures), yogurt, applesauce, baby food, or other smooth foods. Do not offer regular feed or access to grass and other large foods just yet. Take note of the bird's droppings and the amount of soft food and liquids she ingests. You may repeat this regimen for a third day. When the size of the crop reduces, slowly reintegrate harder foods until the bird is eating her daily crumble, pellet, or feed again, and reintroduce her to the flock.

If the oil application fails to move the blockage, the last resort would be to manually remove the blockage by encouraging the chicken to vomit. An Internet search will produce dozens of instructions on how to help the bird induce vomiting to clear the crop. However, this technique is not recommended for the novice chicken keeper. Instead, my recommendation is to take the bird to a vet and allow the skilled hands of a professional to help.

Marek's Disease

What It Is

This highly contagious virus is caused by the herpesvirus—six different herpesviruses in fact. It's incredibly common, and actually, most birds already carry the dormant virus, just waiting for the right conditions to emerge. Indeed, Marek's disease kills more chickens than any other disease, and unfortunately, it's not very pretty. There are both conventional and natural ways to prevent it, each with varying results.

Marek's disease (MD, or, simply Marek's) is also referred to by the names neuritis, neurolymphomatosis, or simply range paralysis, due to its effects on the bird's movement. The incubation period is only about two weeks, and once infected, it is nearly 100 percent fatal—so there is no cure once onset is detected. Nearly half of all unvaccinated flocks become afflicted with Marek's disease, although the vaccine (discussed later) is not foolproof. Five percent of vaccinated birds still get it.

Marek's is highly contagious and airborne, so once it starts to spread, it's not easily controlled. It is more common in standard birds than in bantams, and all infected birds, whether showing symptoms or not, shed the virus, contaminating the coop, run, pen, enclosure, or pasture. This may not necessarily be a negative attribute, however, as birds gradually exposed to the virus may develop natural immunity under ideal circumstances.

What It Looks Like

Marek's is ugly. It's not fun to see any of your birds suffering from it, and it's also quite painful for an afflicted bird. Marek's can afflict a bird of any age. The most obvious symptom is the progressive paralysis of the legs and wings. A bird with Marek's will be unable to move, will struggle to walk (in the early stages), and will appear crumpled on the ground. The infected chicken may show signs of weight loss, labored breathing, or diarrhea. Eyes may appear gray, look sunken, or pupils may be irregularly shaped. Blindness may also occur. Wings may look droopy, and if the bird is unable to move, others may peck it. Sudden death is also common.

The leg paralysis of Marek's may also be mistaken for the leg weakness of botulism poisoning, rickets, and cage fatigue, all of which are not very common. Considering the

relatively high prevalence of Marek's, if you notice your bird displaying these symptoms, consider that it may have Marek's disease first.

What You Can Do about It

Though Marek's disease may appear in a chicken of any age, the first few weeks of life have the highest chance of infection.

Prevention. Marek's just may be one of the nastiest diseases that may befall your flock—and one of the most important diseases with which to take preventive care. Before getting a flock of chickens, either chicks or adults, taking steps to prevent Marek's in your flock is an absolute necessity. Here are a few options.

Vaccination is the most common and easiest way to prevent Marek's in your flock. Derived from the turkey virus, the live vaccine must be administered before the bird's exposure to the Marek's virus, which is why vaccinating day-old chicks is common practice and often recommended. If hatching your own chicks, administer the vaccine as soon as possible after hatching and keep all chicks in isolation for immunity to develop over the course of a week. Keep in mind that a vaccinated chicken may still become infected with Marek's—the vaccine is effective in about 95 percent of birds.

Natural immunity is possible through gradual exposure to the virus against the disease. But keep in mind that this method can be ineffective if the young bird is exposed to too many of the microbes before its immune systems are mature enough to handle it. It's a delicate dance. Natural breeders prefer this method because it's a fairly easy selection tool: Cull the birds that develop the illness and continue to breed from those that are strong and vital. The philosophy here is that vaccinating the *entire* flock masks the weaker individuals that succumb to the disease, weakening your breeding program and the health and strength of your best birds in the process. Another way to develop natural immunity in a flock is to raise a few turkeys with your chickens. Turkeys are host to a similar virus (harmless to chickens) that keeps the Marek's virus at bay. If you choose this route, educate yourself about raising turkeys with chickens, since it can cause other health issues if not done properly.

Stress reduction and the related care that comes with it are a surefire way to keep immunity boosted and Marek's at bay. If your coop is overcrowded, expand. If you're considering a move, move your birds at night when you can easily corral them and transport them to and from a moving crate or box.

Boost immunity by providing the probiotic supplements mentioned earlier. Make sure they always have fresh water, age-appropriate feed, and supplements such as grit and oyster shells for laying hens at all times.

Treatment. If one or more of your birds does become infected, be prepared to take the bird to your vet as soon as possible to confirm the diagnosis.

Newcastle Disease

What It Is

Newcastle disease is a highly contagious viral respiratory disease that affects all poultry species. If one member of your flock is infected, chances are high that the disease will rapidly spread throughout the flock (if it hasn't by the time you diagnose it). Thankfully, death is not guaranteed

with Newcastle disease, since mortality rates vary widely, but it is rather common. Many chickens can and do survive and subsequently become immune, although they can remain carriers for some time afterward.

A flock's survival depends on the health of each of its members, so it's in an individual's best interest to stay healthy and appear healthy, even when sick. For this reason, take the situation seriously when you see an ill or injured chicken, such as this hen with Newcastle disease.

What It Looks Like

As with other avian respiratory diseases, symptoms include gasping, wheezing, coughing, and general difficulty with breathing. However, Newcastle disease differs from respiratory infections in that it affects the nervous system as well. Muscle tremors may occur, and if left to progress, eventually paralysis (you would see drooping wings, twisted legs, and other deformities). If infected hens continue to lay, their eggs will likely have deformities as well, with calcium deposits and other out-of-the-ordinary markers. Green, watery droppings may also be a symptom but, of course, are not the sole indicator. Newcastle disease shows up quickly and makes its way through a flock fast, in about a week's time.

What You Can Do about It

A vaccine is available and many large-scale operations employ its use, but it is only effective against several strains of the disease, and you should consider vaccination only if it is prevalent in your area. Unfortunately, good management practices do little to ward off Newcastle disease. The best method of natural prevention is to breed resistance in your flock, or purchase from a breeder who raises strong, healthy birds.

Pasted Vent (Pasty Butt)

What It Is

Pasted vents are rather common in very young chicks and although easy to remedy, can be fatal if left untreated. Pasty butts occur when feces dry and seal the vent, preventing the chick from discharging waste. Pasted vents are usually most prevalent in the first two weeks of life.

What It Looks Like

Pasted vents aren't pretty, but as far as chicken ailments go, they're easier to stomach than most. If any of your chicks has pasted vents, you'll notice manure dried on the vent and perhaps stuck to the surrounding down feathers.

What You Can Do about It

Monitor your young chicks multiple times per day, particularly if you have purchased them from a hatchery or breeder, and they have spent some time in transit. You'll need to monitor their feed and water intake as well. Check each chick's rear end at least twice daily. Should you find a pasted vent, remove the afflicted chick right away (mostly so you remember which one it is). With a warm, wet washcloth or towel, gently remove the dried waste from the vent and surrounding feathers. Repeat as necessary. Be careful; young chicks have very delicate skin, and pulling at the down feathers can hurt. The pasted chick may squawk in protest, but rest assured, it must be done. If you don't fix the issue right away, the chick won't survive.

A pasted vent can quickly kill young chicks.

Prolapsed Oviduct (Blowout) and Egg Binding

What They Are

Prolapsed Oviduct. When an egg is completely formed and ready to be laid, the oviduct's uterus pushes it out through the cloaca, turning inside out temporarily in order to complete the egg's exit from the oviduct. Called prolapse, this is a natural part of egg laying. In a normal, healthy bird, the uterus immediately retracts into the vent. When the uterine tissue remains prolapsed outside of the vent, however, it becomes the serious condition known as having a prolapsed oviduct, known casually with chicken-folk as blowout.

Egg Binding. Another reproductive affliction common in a laying hen, egg binding is often confused with blowout. While some of the symptoms and repercussions are similar, the two maladies are actually quite different. Egg binding occurs when an egg remains lodged in the hen's reproductive tract and she is unable to pass it, accumulating other eggs behind it and rendering her unable to defecate (or lay eggs for that matter). It is fatal if the hen cannot pass the egg.

The causes of both reproductive ailments can be related to keeper management practices, or they may be entirely out of your control. Very young layers or hens laying extra-large eggs are generally more prone to both issues. Very sick hens, hens suffering from a heavy worm infestation, or hens with scar tissue from a previous binding or prolapse affliction are more susceptible to these predicaments. Overweight hens that are fed a fatty or inappropriate diet and not permitted to move about may accumulate fat stores around their reproductive system, also causing complications in the laying process. Finally, these two issues may also be related to genetics: Some breeds are developed for the highest possible egg production, and their bodies may develop eggs that are too large for them to pass.

What They Look Like

Prolapsed Oviduct. Blowout is most commonly "diagnosed" by the characteristic presence of tissue outside of the vent. In extreme cases, the hen may even have internal organs exposed through the vent. In cases where a prolapsed oviduct has been neglected, there may be blood and/or bloody tissue present due to picking from the other birds. More than likely, the afflicted hen will have reduced or ceased laying entirely and is being cannibalized by the flock. Pecking or picking at the exposed tissue around the vent by other birds is a very common (and serious) indicator that something is wrong and requiring a closer look.

Egg Binding. Because egg binding is internal by nature, it necessitates a closer inspection to diagnose. Hens suffering from egg binding will usually demonstrate some or all of the following behaviors:

- Straining to lay: The hen's vent may pulsate, and the tail may pump up and down.
- Immobility: sitting on the ground, standing very still, or not roosting. A lodged egg can put pressure on the bird's spine, causing paralysis, making movement hard and painful.
- Lack of eating, drinking, or relieving herself.
- Taking the odd posture that I call "the penguin" stance: The bird shifts her body upright, dips her tail toward the ground, and puffs out her feathers, remaining still.

What You Can Do about Them

Prevention: Prolapsed Oviduct and Egg Binding. First, consider breed selection when building your flock. Choose reputable breeders and avoid strains of high-production breeds that are known to have reproductive trouble. Heritage birds are a great option for the backyard chicken keeper.

Genetics aside, reproductive issues may be the result of a calcium or phosphorous deficiency, so it is important to provide proper nutrition to your flock. Stick to a diet of organic feed appropriate to the age and life stage of your birds. (See chapter 7 for everything you need to know about commercial feeds.) Grit and oyster shell, of course, are always a must.

With that said, even the best commercial or homemade feed will not prevent illness if your birds are confined to small spaces and forced to be inactive. Best management practices for the small flock include free-range time to move about, stretch, dust-bathe, scratch, peck, and forage. If free ranging on pasture isn't a possibility for your flock, provide food scraps and other stimulating treats to keep your birds active in their enclosure. (See chapter 7 for great ideas and recipes for homemade treats.) Finally, create a space for your birds to dust-bathe; this is the ground fowl's natural form of keeping clean and an important social activity for the flock.

Allowing your charges to engage in these natural chicken behaviors will keep your hens healthy, fit, and laying safely.

Treatment: Prolapsed Oviduct. If caught early, a prolapsed vent may be treated with a hemorrhoid cream applied topically to the vent. These creams (the same kind used by humans) may be found at any local drugstore. As with most chicken ailments, isolate the afflicted hen as soon as possible to prevent bullying and pecking from the flock. (The cannibalism inflicted on an injured hen can be brutal and ruthless and may result in the other birds pulling out sensitive tissues and internal organs, leading to shock and ultimately, the death of the hen.)

While in isolation, don a glove and continue to apply the hemorrhoid cream, gently pushing the exposed tissue back into the vent. Be persistent; it may take three or four days of application to make a difference. To prevent further reproductive issues during healing, attempt to reduce the hen's frequency of egg laying by limiting her light exposure. Keep her in a coop, barn, shed, basement, or other room with low lighting. Aim for a light exposure of less than eight or nine hours maximum per day. Keep the prolapsed hen isolated until she improves and the vent returns to normal.

If the hen's vent does not return to normal, you will have a tough decision to make at the vet's office. Either the professionals will be able to handle the situation, or you may choose to have the bird humanely euthanized. If the hen lives, it would be prudent to remove her (and any other blowout- or binding-prone birds) from production and permanently from your breeding program. Doing this will eliminate the chances of her passing along the trait to future generations of your flock.

Deciding When to Visit a Vet

Hopefully, prior to getting your chickens, you established a relationship with an avian vet who is already familiar with you and your flock. If at any time you feel uncomfortable caring for a chicken or if you feel that a situation or injury is out of your realm of experience, call your veterinarian. I grew up around many animals and have a great intuitive sense around most mammals, but chickens are different. It has taken years of experience with many chickens to be able to "read" my birds, determine how sick they are, and assess whether my skills are adequate to help them. Chickens tend to hide their illness or injury and can go downhill fast. Don't waste time on feeling upset that you can't do much to help them (there may not be anything you can do). Go ahead and call your vet sooner rather than later.

Note: Keep in mind that not all vent picking is due to blowout. If you have flock members with severely picked or consistently bloody vents, take a look at your coop's setup to see if any birds are inadvertently roosting, eating, drinking, or laying above or in front of other members of the flock. This is an occasion where simple flock management and attentive stewardship can make a world of difference.

Treatment: Egg Binding. Again, the first step is to isolate your sick hen. In this case, a calm, quiet, and dark environment may be enough for her to pass the egg on her own. The dark, mimicking the shortened days of winter, signals her system to reduce the production of eggs, and the isolation eliminates the possibility of cannibalism she may encounter from flock mates who notice her condition.

Gently feel the hen's abdomen for the lodged egg. Next, check the hen's vent to see if the egg is lodged within reach. If it is, you may be able to take matters into your own hands, so to speak: Don a glove and utilize vegetable oil, mineral oil, or Vaseline to extract the egg. Handle her with great care as you do this; you don't want to break the egg since a broken egg in the oviduct can lead to infection. If you can, get someone to help you with this task.

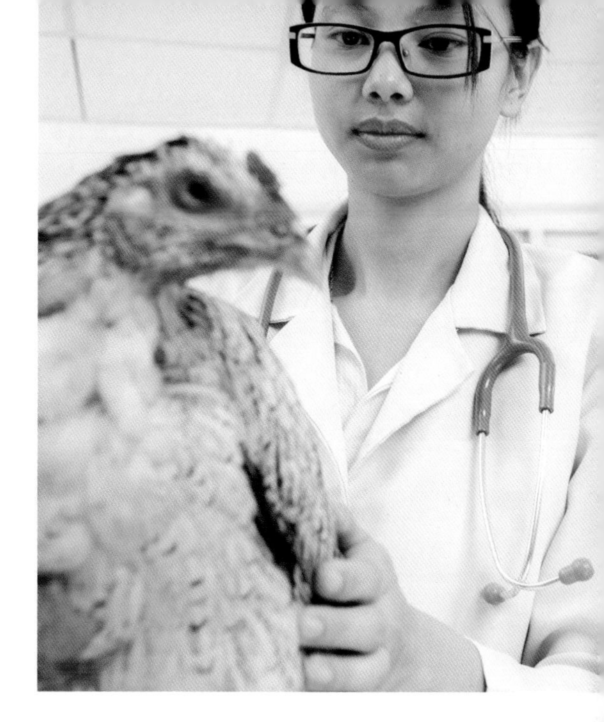

It's a good idea to establish a relationship with a reputable avian veterinarian in your area before your flock gets sick.

If isolation doesn't work, the most successful way to bring a hen back from egg binding is to provide a warm bath or moist heat. Here's how to go about doing that.

In a warm (not hot) water bath, submerge the hen's bottom for at least 30 minutes. While it may seem like a long time to hold a chicken in water, submerging the hen's bottom will help to relax the vent area and facilitate the passing of the egg. There's a small catch, though: Chickens don't typically care for water. Speak gently, move slowly, and massage her belly in the direction of the vent; she should calm down and begin to relax. This method can be done several times a day over the course of a few days if needed.

With the same concept in mind, an alternative way to encourage the egg to move is to utilize the following method using moist heat:

1. Isolate the hen in a wire-bottom cage or kennel with a basin of steamy water beneath it. (The water should be warm enough to provide steam but not hot enough to burn.)
2. Provide a heat lamp and cover the makeshift sick bay with a sheet or towel to retain the moisture.
3. Monitor the heat with a thermometer and keep the temperature warm, but do not exceed 102°F (39°C).
4. As always, provide fresh, clean drinking water.

Is It Blowout or Binding?

You know something's wrong in the egg department, but is it blowout or binding? When you see a distressed hen and suspect a reproductive issue, take a closer look at your bird and use the table below to narrow the possibilities. When in doubt, isolate the hen, provide her with food and water, and call your veterinarian.

Description	Prolapsed Oviduct	Egg Binding
A decrease or cessation in laying	X	X
A lack of appetite	X	X
Puffed-up feathers, "penguin" stance		X
Standing still or lack of regular movement		X
Discolored or unusual vent	X	
Tissue protruding from vent	X	
Often afflicts young pullets	X	X
May result in the injured hen becoming a victim of bullying/picking/pecking	X	X
May result from poor/inappropriate nutrition	X	X
May result from attempting to lay eggs that are too large	X	X
May be due to a calcium or phosphorous deficiency	X	X
May be fatal	X	X

The moist heat method should work within a few hours. If you see an egg, she has laid the culprit and will show symptoms of returning to normal.

If nothing seems to be working and your hen is deteriorating, take her to the vet immediately. There will be options and choices to make. A trip to the vet will likely result in a calcium gluconate injection (which may help your bird expel the egg) plus a round of antibiotics. An alternative is a costly x-ray and possibly surgery, resulting in a hysterectomy. If the case is severe, euthanasia may ultimately be the most humane option. Do not hesitate to make this decision. Egg binding can be incredibly painful for your hen.

Safe Handling Techniques for Chickens and Wranglers

Chickens are squirmy, flappy, awkward creatures to hold. They have sharp beaks, curved talons, and long, outstretched wings that can inflict all manner of discomfort on their caretakers if improperly handled. Proper chicken-holding etiquette is a must, especially if you're working with children or you regularly take your birds out in public. Learning how to hold a chicken (and practicing and becoming comfortable with it) is a boon to the handler who wants to stay ahead of disease and be attentive to injuries.

It's important to understand that chickens do not really *want* to be held. This is less a reflection of their affection for you and more about the fact that they are naturally prey animals. To be held (in their minds) is to be "caught," indeed a scary notion for a chicken.

For this reason, it's important to hold a chicken securely when you do pick her up. Use a firm and confident grasp, but don't squeeze (this is an important distinction to make for very young children).

There are dozens of ways to hold a chicken, and every keeper has his or her style. The "proper" way, though, ensures a safe bird and a safe keeper, with minimal stress to both. Here's how to do it:

1. With the bird's head and beak facing you, place the bird's breast in the crook of your elbow. Your forearm should run along the underside of the bird.
2. Hold the bird's two feet between your fingers: one leg between your pinky and ring finger and the other leg between your pointer and middle finger.
3. Place your free hand on the bird's back, extending your hand to hold each wing down gently. Keep the wings close to her body to help her feel more at ease.
4. Hold the bird close to your body, securely but not tightly.

This handling technique helps the bird feel safe without compromising its vision or disorienting it. Practice this technique with each of your chickens regularly. When your birds become comfortable and familiar with your handling them in this way, you'll find you can easily inspect them for injury, weight loss or gain, and other health concerns by using your top hand. The best part of this handling technique is that it keeps the feet and the vent (arguably the two nastiest parts of a chicken) safely facing *away* from you.

Each chicken keeper has his or her own style for handling and holding chickens—and some birds are more tolerant than others.

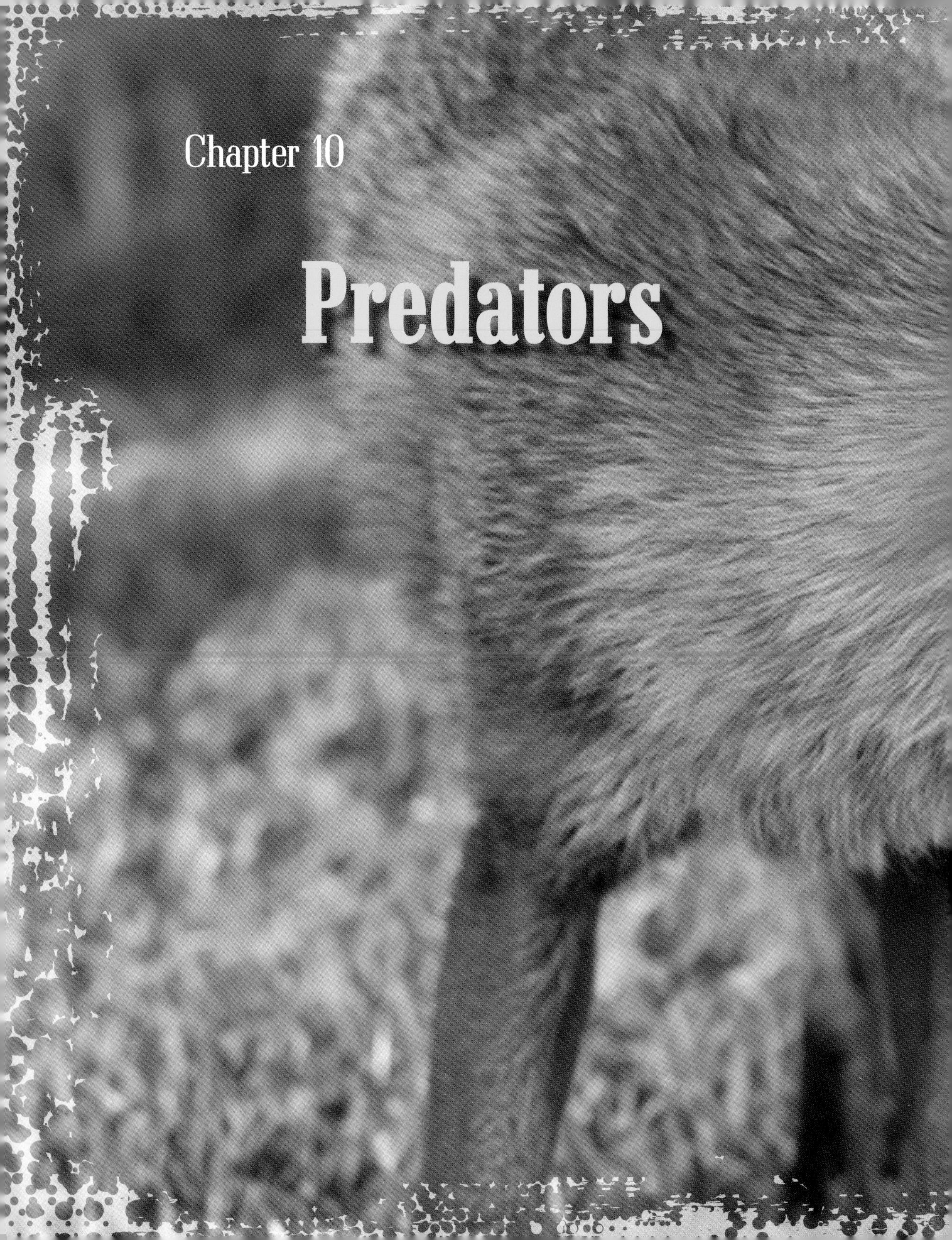

Chapter 10

Predators

Chicken predators come in all shapes, sizes, and species. They fly, crawl, walk, stalk, and slither. Some gain access to your birds by climbing walls, others by slipping through fences, some by digging under enclosure perimeters, and a few by simply charging in the light of day. Some are big. Some are small. Some are so crafty they can pass undetected until they strike.

Chickens are rarely safe, and they know it. By nature, chickens tend to be standoffish, skittish, flighty, and a tad bit paranoid. Chicken fanciers often find their behavior quirky and endearing, but what we see as "just being weird" is really a well-honed defense mechanism. It's no wonder chickens are constantly on the alert. Nearly every predatory creature, domestic or feral, finds them to be easy, tasty prey. On top of that, chickens have very few natural defenses. They have poor eyesight in low light and no teeth, claws, or strength with which to defend themselves. Between the natural fight-or-flight response, flight is the only viable option—and they can't even do that very well.

So, as their keeper, it's your duty to use your wits, tools, and resources to ensure the flock's safety. The easiest way to do this is by working *with* their natural instincts and the tools they *do* have in three ways. The first defense, of course, is by building and maintaining a secure coop. The second is by locking up behind your birds each evening, making sure they're safe at the most vulnerable time of day. The third is by thinking like a predator. The latter is the best way to stay one step ahead of the marauders and to truly keep your birds safe. What are your local predator's strengths? How do they gain access to chickens? How would they maim or kill, and what time of day do they tend to strike? These are all important questions any chicken keeper must ask about the predators that hunt nearby. This chapter goes inside the minds and instincts of the most common North American chicken predators and explains how best to protect your flock from their advances.

Reading the Clues

Chickens are so deceptively easy to care for and have so few needs compared to other pets and livestock that it's sometimes easy to forget that they are one of the most vulnerable. It's easy to become complacent in the daily routine and let your guard down, even just once. That one slip-up—the *one*, tiny gap in fencing or the *one* time you forget to lock them up at night—could be a predator's way in and spell disaster for your flock. It's critically important to take the necessary precautions and establish a good defense from the get-go. Don't wait until a predator has already paid your flock a visit.

Check Your Chicks

Young chicks make easy pickings for a number of predators, especially if they are housed outdoors with an adult flock or with a broody hen. Due to their small size, constant peeping, and obvious vulnerability, the chicks' presence alone may be enough to attract nearly any predator featured in this chapter. Some animals, such as cats or skunks, would rarely pay much mind to a flock of adult hens if chicks weren't present. So, if you do brood your chicks outdoors or allow a mother hen to care for her own little ones outside, take extra care to protect them with diligent daily checks, a secure dwelling, and nightly lockups.

If and when an attack happens, be prepared to don your detective's cap. Unless you catch the marauder in the act, you'll be relying on clues at the crime scene to determine which species made the attack. It can be surprisingly hard to figure out who was responsible. Ask yourself the following questions as you assess the scene. Then, read "How to Determine the Culprit" below to discover which animal's kill style best matches what you see.

1. Check for obvious points of entry. What do you see as you scan the coop and run's perimeter? Are there gaps or torn holes in the fencing? Signs of digging? A window or door left ajar or pried open?
2. Check for obvious animal tracks around the enclosure. If you have muddy or snowy conditions, you may get lucky and find some.
3. How many birds were killed?
4. What time of day did the attack happen?
5. Were any birds eaten? If so, which body parts?
6. Are there any missing birds?
7. If there are surviving birds, what is the nature of their wounds?

How to Determine the Culprit

Chickens are easy prey for so many predators that it's hard to keep track. From feral cats to foxes and hawks, each predator has its own distinctive modus operandi that serves as a calling card, providing clues to what you're dealing with. A description of the most common of these chicken stalkers is described below.

Coyotes

Coyotes are wiry, savvy members of the dog family that are rather widespread across the American landscape. Once residents of the plains and deserts, these crafty canines have keenly adapted to modern infrastructure and are now found in nearly every state and city in North America. Coyotes are scrappy omnivores who eat nearly anything—rodents, rabbits, snakes, small livestock, fruit, neighborhood garbage, and, of course, your chickens.

Coyotes are smaller than wolves and can be mistaken for domestic dogs, albeit on the skinny side. They're very clever, as Native American folklore suggests, and they

Coyotes are smart hunters that often travel solo or in pairs.

tend to form packs in the winter for easy hunting. However, it's rather common to see individuals wandering and hunting solo or in pairs. Coyotes have a sharp sense of smell, great vision, and can run up to 40 miles (64 km) per hour.

Calling Card

Even if you live in a densely populated suburban neighborhood, don't rule out a coyote in the event of an attack on your flock. Truth be told, it can be rather difficult to distinguish between a coyote attack, a fox attack, or a wolf attack. Your first clue, of course, will be which animals likely reside in your region of the country. If you notice some of the signs below following an attack, a coyote could be considered the culprit:

- Missing chickens.
- Scattered feathers.
- Very few clues of an attack.
- Early morning attack (although coyotes may strike either day or night).
- Weak, old, sick, or slow birds taken first.
- Broken necks. (Coyotes and other canines prefer to break the neck of the prey first, but they will grab any part of the chicken they can reach and make off with the whole bird.)

Your Flock's Defenses

Coyotes are known for digging under fences as well as scaling over them. The best defense for outdoor runs and enclosures is to erect tall, strong fences and bury heavy-duty wire mesh at least 1 ft. (30 cm) into the ground around the run's perimeter. Hardware cloth, not chicken wire, is recommended. If you prefer not to dig, you may instead choose to fan out the wire mesh (also about a length of 1 ft.) in an apron around the base of the

enclosure's fencing. Electric net fencing (the kind with smaller openings, rather than three continuous strands) also works well to protect pastured birds from coyotes if used intermittently. If used daily, coyotes may learn ways to get around the fencing, jump over it, or learn the times of day when it's not "hot."

Domestic and Feral Cats

Domestic and feral cats appear smaller than most standard-sized chickens, but don't be fooled. Cats may consume young chicks in their entirety or attack larger birds that are sick, injured, isolated, or unaware. Of all the predators on this list, cats pose the least threat to a flock of adult chickens, but this doesn't mean you should let your guard down.

Once in a great while, a very hungry feral cat, a large housecat, or a particularly bold feline will take on a grown chicken. If the chicken is a bantam or is rather petite, the cat has a better chance of causing some real damage. However, the true danger that cats pose is really toward small chicks. Housecats, in particular, find squeaking, chirping, flying chicks to be alluring "play things," and they won't waste time inflicting fatal wounds.

Calling Card

Domestic and feral cats hunt for both food and sport. The remains from a cat attack may indicate either or both scenarios. Consider a cat to be the culprit if you find the following at the scene:

- Dead chicks in a brooder (indoors).
- Dead or fatally wounded chicks around the house (taken from the brooder and played with).
- A large mess; lots of feathers and bird parts strewn about.
- Muscle and meat of the birds consumed; feathers, wings, head, and bones remain.
- Attack occurs either day or night.

Your Flock's Defenses

Since the vast majority of cat attacks occur in household brooders, the remedy is simple: Secure vulnerable chicks away from housecats. Always. A brooder with a mesh top and solid sides is all well and good, but it may not keep a determined feline from getting in. Cats especially like to paw at their prey and will easily slip a leg through chicken wire to investigate the chicks. Remember also that cats can climb and jump, and they like to pounce. Ultimately, the best line of defense is a closed door. Keep your chicks in a separate room or part of the house away from the cats entirely. If you are keeping a brooder in a barn, shed, or other structure where a mouser lives, consider building a partition or moving the chicks to a more secure area.

Domestic and Feral Dogs

Domestic dogs truly account for the majority of backyard chicken losses. They are abundant in any neighborhood or city (where there are people, there are pooches) and are often the most overlooked predator because of their status as pets. Any dog, no matter its size, *may* pose a threat to chickens. Please note that this does not mean that all dogs

Great Pyrenees mountain dogs and other guardians are bred to protect poultry and other livestock from a variety of predators.

do pose a threat. In fact, there are many individual dogs and various breeds that make fabulous livestock guardians and live quite amiably with poultry.

Other than these few trustworthy canines, you should exercise precaution around new or unfamiliar dogs, whether they have a great reputation or not. Domestic dogs with a high prey drive will kill for sport; feral dogs, on the other hand, are more likely to consume their prey. A feral dog's behavior is more like that of a coyote, so if you suspect a wild canine may be the culprit, be sure to read the section on coyotes, starting on page 163.

Contrary to how the scene unfolds, many dogs that attack chickens do so out of play, rather than with the intention to kill. There are also some dogs that will chase chickens relentlessly, forcing them into dangerous situations, causing injury, heart attack, and/or death, without meaning to. It's of the utmost importance to note that there is not one breed of dog that is any more dangerous than another—*any* dog with a strong prey drive may find chickens to be irresistible playthings. Be wary of any domesticated dog that is not trained or adapted to life with chickens, no matter its size.

Calling Card

Unlike most of the other predators in this chapter, domesticated dogs will often strike during the day. Dogs usually continue their killing spree until all of the birds in a flock are dead; however, you may find some that have survived a dog attack. These birds will likely be fatally wounded and should be humanely dispatched as quickly as possible. Since many dogs that attack chickens kill for sport, the telltale signs of a canine presence can be pretty clear:

- Most, if not all, of the chickens dead or fatally injured.
- Bodies scattered around the enclosure haphazardly.
- Chickens with broken necks.
- A very big mess: Blood and feathers everywhere. Surviving chickens may have large puncture wounds, broken legs or wings, or skin pulled off.
- Chickens mauled but not eaten.
- Torn fencing where the dog(s) has gained access.
- Holes dug under fencing where the dog(s) has gained access.
- Whole eggs missing or empty shells in and around the nests.

If a dog attacks your flock, it's important to remember a few things. First, it is not the dog's fault. It may be painful for you or the dog's owner to accept responsibility for what happened, but remember that the dog was simply acting on instinct. As the flock's keeper, it's your responsibility to keep your birds safe. It's the dog owner's responsibility to monitor his or her pets and keep them on leash or in their own yard or home. If the attack occurs by another person's dog, rather than your own, report the incident to your municipality's animal control department. While rules vary from city to city, the dog's owner is likely responsible for reimbursing the cost of your lost birds, damaged fencing, and other financial losses you may have incurred.

Finally, if any of your chickens should survive a dog attack with little apparent physical injury, remember that they will still be suffering from the severe stress and trauma of the event. They'll likely cease laying for a while, from a few days to several weeks. Watch them closely for other signs of distress and minimize stressors in and around the coop for several weeks following the attack.

Your Flock's Defenses

Protect against dog attacks in much the same way you would against coyotes: Erect tall, secure fencing, bury fencing or mesh along perimeters, or use electric netting for pastured birds.

Do exercise a few extra precautions when protecting your birds against man's best friend. If you own one, first determine if your own dog is trustworthy around poultry. Stay on top of his training and be consistent with it. If need be, take measures to keep your pup and your poultry separate in order to keep them safe. Next, speak with neighbors about their dogs and any stray dogs you encounter in your neighborhood. Do your neighbors' dogs live primarily indoors or outdoors? Are pet dogs permitted to roam the neighborhood? Do you often find stray dogs in your neighborhood? Share your concerns with neighbors in order to reach an agreement and report stray animals to your municipality's animal control division.

Many are surprised to learn that the domestic dog is the primary predator of backyard chickens.

Foxes

The fox is the quintessential chicken killer. Sly, smart, and savvy, foxes will assess an area before striking and will do so only after they know the coast is clear of humans and other guardians (such as guard dogs). Foxes prefer to make their moves in the early morning hours or evening, but they've been known to attack during the day as well.

There are four types of foxes found in North America: the red, gray, arctic, and kit—with the red fox being most common. The four types vary slightly in appearance, locale, and behavior, but they are collectively considered the smallest wild dog on the North American continent. What they all have in common is their cunning: Foxes are

The fox is a classic chicken predator. To defend against these predators, keep grass short, fences tall, and employ a livestock guardian while your flock free-ranges.

incredibly bright and learn quickly; they pick up on patterns, such as when you're at home and when you're not. This makes them challenging adversaries to outwit when they're set on having your chickens for dinner.

Luckily for most backyard chicken keepers, foxes rarely visit urban and suburban neighborhoods. Of course, there are always exceptions. If your home is situated near dense forest, sprawling land, or if you have a large amount of property, it's possible that a vixen will make her den nearby and discover your flock for what they are: easy pickings. Foxes are rather territorial, too, so if you've confirmed that there are any residing nearby, they'll likely be there for a while.

With red foxes populating every state except Florida, it's better to be prepared for an attack rather than caught off guard. Foxes can climb, but they prefer to dig under fences or attack birds while they're ranging. They will bide their time, stalk the birds, and strike when the hens are far from safety. Red foxes can also swim, run up to 30 miles (48 km) per hour, and they're also excellent jumpers, capable of pouncing up to 15 ft. (4.5 m) in the air.

Calling Card

Foxes have partially retractable claws that allow them to quietly sneak up to prey and reveal their talons for a quick and efficient capture. Though this predator is part of the canine family, the fox's hunting style is more akin to that of a cat. She stalks prey, makes her move by running quickly or pouncing, and then uses her sharp claws to pin down prey. If a fox attacks a flock of chickens while they are free-ranging, she'll likely grab only one, maybe two, at a time. If she gains access to a henhouse, the fox will kill and carry out as many as she can. Suspect the fox if you see some of the following clues:

- One or two missing chickens.
- Chickens disappearing while free-ranging.
- Little or no blood.
- Missing bird(s) with no evidence other than a few clumps of feathers.
- Missing bird(s) with no evidence at all.
- Attack occurs in the early morning or evening hours.
- If the attack occurred in the coop, several birds missing and others injured.
- Injured survivors likely to have deep neck or back puncture wounds.

Your Flock's Defenses

Mow grass regularly and keep brush cut back to reduce the cover that foxes use while hunting. If your birds pasture some distance away from your home or any other regularly occupied human dwelling, make use of tall fencing with a buried perimeter or enlist a guardian, such as a livestock guard dog, to protect your flock. Electric fencing does little to deter foxes; they'll either slip between the lines or jump over it.

If foxes frequent your area or you know there's a den nearby, pull out all the stops. Bury hardware cloth perimeters around the coop and run and lock up your birds each night when they retire for the day. Let your birds free-range with company only. Leave a chicken-friendly dog outside or be outside to watch them. Remember, foxes are clever, and they'll strike when your guard is down. Once they've had a taste of chicken, they'll be back for more, again and again.

Mustelids: Weasels, Minks, Ferrets, Fishers, and Martens

Weasels, minks, ferrets, fishers, and martens are just a few among the small carnivorous mammals considered part of the mustelid family, commonly called the weasel family. These aggressive hunters are usually recognized by their long bodies, short legs, rounded ears, and five-toed feet with non-retractable claws. If you've never seen the damage they can do to a flock of chickens, you would almost think they were cute. Animals in the mustelid family tend to smell rather pungent as well; their powerful anal scent glands release a persuasive repellent odor. These little carnivores are nosy by nature, very active, and constantly moving around on the hunt for prey.

As with the other predators in this chapter, whether or not your flock is vulnerable to these carnivores depends on your location. Since most of the hunters in this animal family are rather small, chickens are not usually their first prey of choice. In fact, many members of the weasel family that live in North America prefer to hunt small animals, such as mice, rats, and voles. Because of their svelte frames, these

Tales from the Coop

Several years after our first predator attack with a fox, we experienced firsthand how stealthy an attack from above could be. Our third rooster, a stately but stout salmon Faverolles fellow named François, watched over our flock. The third time must be the charm, because he was easily the best rooster we'd ever had. Where others had been aggressive toward us, François was leery (as a rooster should be) but respectful. He never laid a spur on any of us. Where our other roosters had mated our hens until their backs were bare, François was a gentleman and only provided his "services" sparingly. And where other roosters had failed to protect our hens from predators, François succeeded.

On this particular day, our makeshift guardian dog, Winnie, was inside spending some well-earned time on the couch while we were all out of the house. Errands kept us later than usual, and it was several hours before we returned home. When we checked on the flock later that day, the hens were scattered. Some had dashed into the woods; some had found shelter in the coop. We found François standing very still at the edge of the run, low to the ground, feathers puffed, and blood running down from his comb.

We sprang into action and brought him inside. In the safety of a large dog kennel in the basement, we tended to his torn comb, which had nearly been ripped clean from his head by an attack from a red-tailed hawk. We offered water, supplements, and treated his wounds. We checked on him around-the-clock and whispered words of encouragement. On the morning of the third day, the sound of crowing from the basement woke the whole house, and a wave of relief passed over us. François would live to fight, and protect, another day.

The most common predator in the mustelid family, weasels will steal eggs, decapitate birds of any age or size, and prefer to drink the blood rather than consuming the entire carcass.

little guys can squeeze themselves through surprisingly small holes (about the size of a quarter) in wire mesh and openings in the coop, and they can dig under enclosure walls or climb fences. They're also rather strong for their size. Mustelid hunters are a good incentive to keep your coop clean. If you've encountered any member of the weasel family in your chicken housing, it's likely they were attracted by rodents and decided to stick around to make a second meal of your flock.

Calling Card

Unlike other predators who kill or take one bird at a time, animals in the weasel family tend to kill or injure several birds during one attack. They also prefer to suck the blood of the prey animal, rather than consume large amounts of flesh. Consider that a mustelid mammal may be the culprit if you see some of the following after an attack:

- Chickens killed and collected in small piles (weasel, mink).
- Bites on the back of the head and neck (weasel).
- Only the head or neck eaten or bitten off (weasel, mink).
- Bites around the vent and/or intestines removed or visible (fisher, marten).
- Bodies tucked away to return to later (fisher, marten).
- Small birds, such as chicks and bantams, entirely missing (mink).
- Lingering odor (all mustelids and skunks).

Your Flock's Defenses

Your flock's two main lines of defense against weasel family hunters are, first, to secure small openings and weak points in the coop and second, to keep a tidy coop and storage area. Stop weasels at their point of entry by securing corners and gaps that are larger than a quarter in size. Use hardware cloth with ½ in. (1.3 cm) openings to line windows and to use as fencing in the run. Store feed in predator-safe containers and keep the coop clean of trash, feed bags, and food scraps to reduce or eliminate any rodent populations, thereby not attracting any mustelids. Many of the mustelid predators are cautious around humans, so they'll keep their distance where there is noise and light. Especially practice caution at night by locking up your flock nightly, since these predators are most likely to attack after dark.

Opossums

The opossum, colloquially referred to as a possum, is a silvery-gray nocturnal animal with black feet and ears, a long hairless tail, pointy nose, beady eyes, and sharp teeth. This marsupial—the only one to call North America home—is a great climber, and uses its gripping tail and sharp claws to spend much of its time in trees. A bit of a scavenger, the opossum will eat just about anything, from fruit and nuts to roadkill and nearly anything remotely edible that can be found in a trashcan. When the opossum can't forage, she'll hunt for mice, rats, snakes, and insects, although she prefers to get food the lazy way by scavenging.

Opossums are rather common and live in nearly every corner of the country, in some cities, most rural areas, and nearly everywhere in between. They're famous for their ability to play possum, or play dead, and they'll turn on this act if they feel seriously threatened. Like chickens, opossums are not as dim-witted as they seem, despite the fact that they tend to move slowly and can appear to be rather clumsy. If cornered, opossums may act quite viciously, in fact, so be cautious if you happen upon one. Even though they'll keep mice and rat populations at bay, they shouldn't be allowed to take up residence in a chicken coop or in an outbuilding nearby.

Calling Card

As already mentioned, the opossum prefers to scavenge for food rather than hunting for it. If she finds a way into a chicken coop, she'll go for eggs first, eating them on the spot. Young chicks make easy pickings as well, and if in the mood to hunt, an opossum will target grown chickens as well. Here are a few telltale signs of an opossum intruder:

Opossums may act viciously if cornered; use caution if you encounter one.

- Empty eggshells in or near nests.
- Killed birds or chicks partially eaten on-site.
- Killed birds or chicks with only the abdomen or crop consumed.
- Dead chickens with bites to the neck or breast. (Like members of the weasel family, opossums will sometimes only suck the bird's blood.)
- Dismembered adults or chicks; adults or chicks with broken wings or legs where the culprit reached through large wire mesh and grabbed any body part it could reach.

Your Flock's Defenses

Knowing that the opossum is an adept climber, it's critical to keep fences secure

Raccoons live in nearly every city, town, and suburb in North America, and the damage they can inflict on a flock of backyard chickens can be devastating.

and roofing material flush with the sides of the coop or run. Outdoor runs should have secure roofs. Like some of the animals in the mustelid family, the opossum can squeeze herself through tiny spaces if motivated, so be sure to close off any gaps in the flock's housing or fencing. Use strong fencing with a small wire mesh, as mentioned earlier. Take extra precautions if you house young chicks outdoors. Also, collect eggs daily to remove any temptation. Though they sleep soundly during the day, opossums are very active at night, so locking up your chickens each evening is a must.

Raccoons

Of all the wild predators profiled in this chapter, the raccoon is the most likely to visit your coop and the most likely to do a great deal of damage. For most backyard chicken owners, the raccoon is their birds' main predator, second only to domestic dogs.

Raccoons are characterized by a mask of black fur around the face and eyes, a furry brown body, and a fluffy tail with alternating black and brown rings. Raccoons can grow rather large, topping out anywhere between 15 to 40 lb. (7 to 18 kg), depending on how well they're fed, of course. They have long back legs and short front legs, making them look hunched over when on all fours.

The raccoon is a mammal native to North America and found in nearly every town, city, and suburb across the country. They are both opportunistic foragers and skilled hunters. Raccoons prefer to make their home near a water source, feasting on fish, crawfish, frogs, snails, and other marine life, but they are incredibly adaptable. They'll turn over garbage cans to dine on trash and scavenge for scraps and, if motivated and hungry, may even enter a home or other human dwelling in search of food.

Calling Card

Raccoons are incredibly destructive chicken hunters. These crafty animals will hunt solo or expertly coordinate an attack in a small family group. They're fabulous climbers, and their front paws are deceptively dexterous. They can open latches, undo locks, and open doors. Raccoons are also strong; they can (and will) tear apart chicken netting or wire. When all other attempts fail, they'll thread their arms and hands through small openings and ruthlessly grab at—and tear off—any part of the chicken they can reach. Raccoons are creative and smart and will do anything within their power to get to your birds.

Expect to find quite a scene of carnage following a raccoon attack. Here are a few telltale signs that a raccoon or its family has paid a visit to your flock:

- Dismembered adults or chicks.
- Dead chickens found where they were killed.
- Dead chickens with entrails pulled out.
- Multiple dead birds.
- Dead chickens with missing heads (or the heads of dead chickens dispersed throughout the coop).
- Dead adults with only the crop and/or breast eaten.
- Surviving birds with broken wings or legs (where the raccoon reached through gaps in housing or fencing).
- Surviving birds with head or neck wounds or bites near the vent.
- Chicken body parts (such as legs or heads) or other pieces of torn flesh in or around the water fonts.
- Broken eggs/eggshells in or near the water font.
- Usually whole flock killed, with majority of the bodies left on site (and not carried away).
- Bags of feed torn open and contents dispersed.
- Attack usually occurring at night.

No type of sliding-bolt lock is too much protection against the crafty and dexterous raccoon.

Your Flock's Defenses

Raccoons are determined, savvy, and smart hunters that will get to your flock any way they can. They'll look for (and find) any weak point you've left in housing or security. They'll attack when your guard is down and do terrible damage to a flock in a single night.

Your best defense against these hunters is to provide your flock with the best and safest housing you can afford to build. Because they're skilled climbers, build tall fences. Put roofs on any outdoor runs and close up gaps in housing. It's critically important that you walk the perimeter of your coop, run, and any other enclosure and look closely for any weak points. If they are there, raccoons will find them. When choosing mesh for

The Trap of Trapping

When a wild animal becomes a nightly neighborhood nuisance, a common reaction is to want to remove the animal from the equation in the hopes that their absence solves the problem. For chicken keepers, a predatory wild animal is often more than a nuisance: It can be downright destructive, and the results of its (very natural) actions can be devastating and heartbreaking. Trapping and relocating the animal may seem, to the everyday homeowner, to be a humane, effective, ecological solution to the problem. Unfortunately, it is none of those things.

Trapping and relocating, unless done by a licensed professional, is actually rather cruel and can cause more harm than good. Here are a few of the reasons that trapping and relocating is a bad idea:

- It's ineffective. Another animal of the same species will claim the territory left behind by the animal you removed.
- There are no guarantees you are trapping the animal that has attacked your flock.
- The trapped animal may sustain minor to major injuries, potentially broken bones, claws, wings, or teeth in an attempt to escape the trap.
- Trapping creates orphans. If the animal you capture is a mother with a nest or den nearby, her offspring will starve and die when she doesn't return.
- Relocated animals do not know where sources of food and water are, resulting in starvation and eventually death.
- Relocation disrupts existing wildlife. Wild animals don't just settle in and make new friends wherever they end up. A new animal that has appeared on another's turf will spark territorial disputes in which one or more of the animals sustains injuries or even death.
- If the animal you capture is ill, relocating it may spread disease.
- Some species will attempt to return to their home and familiar territory, dying by traffic or predators along the way.

The most compelling argument against trapping and relocating wild animals is that it simply doesn't work. As mentioned earlier, another animal of the same species will quickly fill any territorial hole left in the ecosystem. If you don't remove or remedy what is attracting the animal in the first place, wild animals will continue to show up.

If that's not incentive enough, trapping and relocating wildlife without a license is illegal in many states. Trapping some species, such as protected birds of prey, could result in hefty fines and time behind bars. Trapping and relocating should be conducted only by trained professionals who are well versed on the individual species, its mating habits, and the time of year, and by those who are meticulous about how they reintegrate the relocated animals. This requires skill and training that most of us don't have. Always contact your county extension office for guidance in finding a wildlife expert to assist you. Laws differ state by state, so be sure to do your research before taking action.

Remember, the best way to protect your flock from unwanted advances by local wildlife is to take the protective measures listed in this chapter. A solid defense is better than an aggressive offense: Batten down the hatches at home and leave the wildlife alone.

your coop, always purchase heavy-duty hardware cloth over chicken wire (which is flimsy and easily torn by a resolute raccoon). As mentioned previously, line windows, doors, and roofs on any outdoor enclosures with ½ in. (1.3 cm) hardware cloth.

Raccoons aren't great diggers, so they will rarely try to dig under fences or coop walls. However, they have an advantage over other predators in one area: the front door. Most importantly, lock up your flock every night. Because of their nimble paws, they can open locks and latches that other predators can't, so it's important to safeguard against these shrewd intruders by firmly fixing knobs, locks, and bolts on any coop doors or windows. Slide-bolt and hook-and-eye latches are too easily popped open by raccoons and should *not* be used in a chicken coop. Instead, use clip latches or spring-loaded hook-and-eye closures (the ones that require an opposable thumb to pull and release). (Refer to page 94 for more on raccoon-safe latches and locks.) When it comes to keeping raccoons out, there's no type of lock that is overkill—it's better to be safe than sorry.

Raptors

Without much cover in which to hide, predatory birds can pose a very real threat to a flock of free-ranging chickens, especially small ones such as bantams, juveniles, and chicks. These birds of prey, which include daytime (or diurnal) predators such as hawks, falcons, and eagles, and nighttime (nocturnal) predators, like owls, are equipped with sharp talons, keen eyesight (or impeccable hearing), and enormous curved beaks designed to expertly tear at flesh. All predator birds, whether they spend their waking hours in the daylight or the moonlight, share a rather stealthy ability to hunt. On the whole, they move quickly and leave little evidence behind—especially when compared to some of the other predators profiled in this chapter.

One of the most common raptors in North America is the red-tailed hawk. It is most easily identified by its broad wingspan and short, fanned tail. Hawks are most obvious to the casual observer when circling overhead, riding ridge currents, or vocalizing to other birds in the area. They will, however, happily find a hunting perch on which to sit quietly and wait patiently for prey—sometimes for hours.

Of the nocturnal predator birds in North America, the common barn owl is the most likely to pay a visit to your chickens or be the cause of any concern to you. Characterized by its flat, heart-shaped face and large, dark eyes, the barn owl uses its immaculate hearing, rather than sight as other raptors do, to locate and capture prey.

The common barn owl is characterized by its heart-shaped face and dark eyes.

Birds of prey pose a significant threat to flocks that free-range without much overhead protection.

While it takes a rather large raptor to carry off an adult, standard chicken (or a very bold and brazen one to make the attempt), do not underestimate the audacity of a hungry raptor. Most predatory birds will home in on small or weak prey, but they are still capable of inflicting damage on any member of your flock with an attempted capture. If your home and coop are located in seasonal hunting grounds for these migratory birds, it would be prudent to plan for any possible threat from above. If you suspect your property or surrounding neighborhood is the stomping grounds for species of predator birds, do your homework by calling your county extension office or enlisting the help of avian experts at your local bird or wildlife sanctuary to learn which birds frequent your area throughout the year. Do not attempt to trap or capture any predatory bird on your own. Many native North American raptors are federally protected and to illegally capture one (that is, without a license to do so) could land you with some serious jail time and unforgiving fines.

Calling Card

A bird of prey is the archetypal hunter. Unlike the other predators in this chapter, true birds of prey are strictly carnivorous—they will rarely forage or scavenge for food to supplement their taste for meat. Some will capture prey and consume it on the spot; others will carry it some distance away to enjoy the meal. If an aerial predator has made a meal out of a chicken, it will begin by eating the breast and organs, tearing clumps of flesh and feathers in a short amount of time. Daytime predatory birds usually take only one chicken at a time, although they may return for more.

Truth be told, raptors rarely leave a calling card. More often than not, chickens that fall prey to raptors will simply disappear; there will be little in the way of tangible evidence to point you in the direction of the perpetrator. Here are a few clues that may suggest a bird of prey is to blame for the loss of a chicken (or two):

- Chickens disappear while free-ranging (usually thanks to hawks, eagles, and falcons).
- Chickens missing, with only scattered feathers remaining (also hawks, eagles, and falcons).
- Surviving chickens have deep puncture wounds (from the talons of—you guessed it—hawks, eagles, or falcons).
- Dead chickens found on the roost at night with missing heads or only the head and neck eaten. Dead or injured chickens will have puncture wounds around the head and neck. (You can blame owls for this one. Once inside, an owl may be moved by blood lust and kill more chickens than it can eat.)
- Piles of feathers under fence posts or similar structures where the raptor has consumed its prey (a habit of nearly all predatory birds).

Your Flock's Defenses

The best way to safeguard your flock from overhead predators is to provide some sort of aerial barrier. Cover outdoor runs with roofs or netting, if possible. Anything that disrupts a raptor's line of sight and ability to maneuver will prevent them from effectively grabbing a member of your flock. Some chicken keepers hang old CDs or DVDs from ropes above outdoor pens to confuse aerial predators (imagine a configuration in the shape of a maypole). The sharp glare from the silver objects in the sun can weaken a raptor's vision during a hunt.

In addition to these measures, provide cover for pastured birds—something they can hide under in the event of an attack, such as low bushes, potted plants, and other vegetation. Even a small structure, such as an old doghouse or shed, is better than leaving a pastured flock completely exposed. Chickens are incredibly perceptive to anything flying overhead and will quickly seek cover if they feel threatened.

If your city or neighborhood allows it, consider adding a rooster to your flock. He will keep one eye overhead for any aerial predators and alert the hens to any danger.

And as always, securely lock up your flock each night. Close any windows or doors that do not have a mesh lining to keep nighttime birds of prey—namely, owls—from getting into the coop.

Rodents

Mice and rats are more of a nuisance than a true threat to adult chickens. While they are certainly capable of killing baby chicks, only a very large, very hungry, and very motivated rodent will attempt to kill an adult chicken. Rats and mice tend to view the coop as a warm, dry shelter in which to make a nest, particularly if there is chicken feed to feast on nearby.

Calling Card

- Missing chicks.
- Partially eaten chicks.
- Missing eggs with no clues.
- Injured chicken(s) with bite marks on its legs.
- Holes chewed into the sides or corners of coop walls.
- Bags of feed that have been chewed and torn open.
- Rodent droppings in and around the coop, nest box, brooder, or storage area.

Your Flock's Defenses

The most important way to keep rodents away is to remove any temptation. Store feed and supplements in airtight containers and don't leave any food scraps lying around. If your chickens don't eat their treats or kitchen scraps within a day or two, remove, clean them up, and throw them away. Keep a clean coop and you'll keep rodents away.

Rats and mice aren't great climbers, so tall fencing will usually deter them; however, they prefer to burrow and chew their way into a shelter anyway. Use sturdy building materials for the coop and for the brooder, especially if it's located in an unoccupied outbuilding (that means no cardboard or flimsy plastic). Be wary of any gaps in your fencing or housing—even a 1 in. (2.5 cm) opening is wide enough for a mouse or rat to slip through. As always, use ½ in. (1.3 cm) hardware cloth mesh to secure any gaps.

Skunks

For a long time, skunks were thought to be related to other carnivorous mammals in the weasel family, but they're actually classified in their own animal family, Mephitidae. Unlike the mammals in the weasel family, skunks rarely hunt grown birds. They tend to go for young chicks or eggs almost exclusively. Fortunately (or unfortunately, as the case may be), the skunk's smell is much stronger than that of the mustelid mammals, so if you are aware of its smell, you can be fairly certain it has paid your flock a visit.

Calling Card

As mentioned above, the skunk's primary calling card is his smell. If you can identify the skunk's smell, along with any of the following signs, chances are great that a skunk was the intruder:

- Empty eggshells.
- Missing chicks.
- Dead chicks.
- Dead chicks with only the abdomen eaten.
- "Skunky" smell.

Your Flock's Defenses

Skunks rarely pose a threat to a flock of adult hens, and preventing skunks is quite simple: If the precautions suggested under any other predator in this chapter are followed,

you probably won't have any issues with skunks. First, keep coops and outbuildings tidy and uncluttered; skunks will be attracted to a coop that offers undisturbed hiding places and shelter. As always, lock up your flock each night. Skunks are nocturnal and will become most active after dark. Finally, keep fences tall and strong. Skunks can't climb, so tall barriers will effectively deter them.

With uncluttered and tidy outbuildings, skunks will rarely pose a threat to your chickens.

Snakes

Like rodents and skunks, snakes are rarely a danger to adult chickens. If they are allowed access to a chicken coop, snakes will happily dine on young birds and eggs but pose little threat to grown birds. While it's possible for some large, exotic snakes to kill and consume whole adult chickens, it would be rather rare. In fact, it's more likely that an adult hen will make a meal out of a snake than the other way around.

Calling Card

Unless you're able to actually catch a glimpse of a snake entering the coop, there will be only a few mysterious signs of his presence:

- Missing chicks; no other clues.
- Whole, missing eggs; no other clues.
- Dead, adult chickens with a wet head (where the snake attempted to swallow it).

Your Flock's Defenses

Snakes prefer to feast on rodents, such as mice and rats, and will take up residence where and when there are ample food sources. That means if you have a resident snake, you likely have a rodent problem as well. To nip the issue in the bud, refer to the "Flock Defenses" section under rodents on page 178. Follow the same precautions for securing the coop, too. Like rodents, snakes are able to fit through very small openings in the coop. Eliminate the food source, and the snake will move on.

Chapter 11

Birds of a Feather

By now, your fluffy chicks have fully feathered into healthy laying hens. If all goes well, you're flush with eggs and have some to spare. At any chance you get, you probably find yourself sharing amusing chicken stories about "the girls" to perfect strangers, while you revel in the surprisingly quaint comfort of keeping chickens. For most backyard keepers, the second six months of their flock's life is a lot of fun. You continue to watch the intricate dynamics within the flock and enjoy their endless antics. The first part of this chapter is for you to learn about how and why chickens behave and communicate the way they do, deepening your understanding of your charges and making you a better steward.

Once you are well versed in chicken social structure, it's time to prepare for routine chores. The second act of this chapter will provide you with a checklist of daily, weekly, monthly, and yearly tasks. Make a copy and hang it where you can see it regularly.

Finally, it's important to recognize that keeping chickens is not for everyone at every point in life. The hope, of course, is that many who take on backyard chicken keeping will see it as a lifestyle change rather than as a hobby. And indeed, for many, it's just that and will lead to years, if not decades, of cohabitation with these fabulous fowl. For others, it will simply not be the right fit. If you've decided to get out of the business of keeping chickens, your reasons are your own. The only thing left to do is responsibly rehome them in the time you have. For those of you needing to find a new home for your chickens, the final third of this chapter is for you.

Flock Hierarchy and Chicken Behavior

A family of chickens, called a flock, is organized by a social structure commonly called a pecking order. Every member of the flock falls somewhere on this spectrum of dominance and submission—no one is left out. The more dominant birds are higher in the pecking order, and the more submissive birds are lower in the order. A flock's hierarchy can be fluid and changes as birds mature or when new birds are added or others removed.

The Pecking Order

In a flock of exclusively hens, there will be one alpha female, and every other individual will be submissive to her in varying degrees. The hen ranked second in the hierarchy will be submissive only to the alpha, the hen ranked third will be submissive only to the second, and so on. In a flock of mixed genders—that is, with roosters, hens, cockerels, and pullets—the males will nearly always be dominant over the females. As pullets mature, they move their way up the pecking order within the population of females. As cockerels mature, they move their way up the pecking order within the population of males, often sparking vicious battles among the mature cocks for the ultimate alpha spot. Old roosters will be challenged more frequently by younger roosters aiming to dethrone them. Fights between cocks can escalate when there are females around. On average, one rooster can adequately watch over, care for, and mate with a flock of about 10 to 15 hens. In very large flocks of mixed genders (30 or more), roosters will naturally create their own small flocks and, for the most part, leave other males alone.

As with any ranked animal social structure, being a top bird has its perks. Those ranked higher in the pecking order have first dibs on food, water sources, the best treats, nesting boxes, dust-bathing spots, and perches, to name a few. Dominant chickens routinely remind the more submissive birds where they rank in the hierarchy by giving warning growls, harsh glares, and the occasional peck.

Maintaining the pecking order can sometimes appear cruel and unforgiving. Some chicken keepers feel compelled to remove or coddle the most submissive birds, but it's important to remember that a balanced pecking order is actually quite harmonious. When each bird knows where she stands in the order, the whole flock is content. Plus, if you were to remove the lowest-ranked bird, the next dominant bird would simply take her place at the bottom of the totem pole; someone must always be lowest in the pecking order.

Tales from the Coop

Our first foray into the harsh world of chicken social dynamics came when one of our ladies broke a nail (no joking!). The presence of blood sent some of the other girls into a frenzy, and it became critical to isolate our injured girl, Sadie. With the bleeding stopped and healing underway, we gave her a few days on her own in the tractor before moving her back into the main coop.

Once re-integrated with the flock, it was clear that everything had changed. Already a rather timid personality, Sadie plummeted in rank among the rest of the girls and took a rather harsh beating. One hen in particular, Madonna, made it her personal mission to keep Sadie from ever coming near the food and water. We gave it time, but the situation did not improve. We decided to come up with a creative solution.

We isolated Madonna in the tractor for several days (we called it a chicken "time out"). When she was re-integrated with the rest of the flock, the wind had been knocked out of her sails, and she did not concern herself with picking on anyone. The hierarchy reestablished itself to a respectful order, and peace was restored.

Keeping the Peace

As keeper, it's important to set the stage for a peaceful flock environment. Here are a few ways to support the social structure and keep the drama to a minimum:

1. Space equals peace. Give your birds enough coop and run space to allow for lower-ranked hens to seek cover and get out of a bully's way, if necessary. (See page 24 for the general recommendation of housing space per bird.)
2. Put out enough feeders and water fonts for everyone, with multiple stations if space allows, so that submissive birds have the opportunity to eat and drink without being bullied or chased away.
3. Reduce bright lights in the coop or brooder. For brooders, use only red heat bulbs, as described in chapter 5. If lighting is used to augment the shorter days of winter for layers, set the lights on

Chicken flocks have a well-defined pecking order. Birds at the top of the pecking order will have first access to the best roosting places.

a timer and give the flock adequate time in the dark to sleep. Daylight can be extended with artificial light for up to 16 hours to encourage continuous egg production without having ill effects on your flock. After that, allow the coop to be dark for the remainder of the night.

4. Be intentional about how and when you introduce new birds to an existing flock. Every time a bird joins or leaves a flock, the pecking order is reestablished. Frequent additions or subtractions can leave the birds stressed from all the reshuffling, which may in turn lead to cannibalism.

Communication

Sit around a chicken coop long enough, and you'll inevitably hear many of the vocalizations chickens use to communicate with one another. Close observation and a bit of free time will give you a better sense of chicken communication than can be described in a book. Even so, there are a few vocalizations to listen for.

The Hen Song. Some hens make a loud noise after laying an egg. This noise is affectionately referred to as a hen song. To many, a hen's song is the quintessential chicken noise: bok, bok, bok,

Chickens quite literally come home to roost each evening at sundown, an activity that is as predictable as the sun rising in the East and setting in the West. As such, the common expression *chickens come home to roost* has implications of justice and inevitability; that is, a wrongdoer must face the consequences of his or her mistakes or misdeeds from the past.

BA-KAK! Once a hen gets going, other girls may join in the chorus, and the song may continue for several minutes. The jury's still out on whether this noise is a victory call at having laid her egg, an announcement to the world, or an exclamation of relief.

The Squawk. A squawk can happen when a chicken is caught off guard, picked up rather roughly, or she is disputing territory or hierarchy. A squawk can also be a call of alarm, as in the case of a predator attack. It's a loud, short vocalization but can continue in the event of an attack.

The Chirp. Chirping and peeping aren't only reserved for young chicks. Hens and roosters alike make soft chirping vocalizations as they communicate with their flock mates.

The Cluck. Nearly all chickens make conversational clucking noises. These noises are often made when the birds are relaxed and contented. Roosters often make soft clucking noises to their hens when they find something tasty to share.

The Croon. Crooning can be soft and sweet (such as when a hen is alerting her chicks to a food source) or a low, warning growl (such as when a broody hen tells you to keep your distance). The meaning of these low guttural sounds is best identified in context. Take note of the other behaviors your chicken is exhibiting while making these noises, and you'll have a better sense of what she is communicating.

Cannibalism

Cannibalistic behaviors among chickens include picking at or pecking other birds and by egg eating. These behaviors range from minor (a few pecks as a reminder of social status) to severe, resulting in irreparable injury and death. Egg eating is a bad habit, certainly, and terribly frustrating for the chicken keeper, but it doesn't compromise the safety and health of a flock as other cannibalistic behaviors do.

Chicken cannibalism has been touched on in other areas of this book, as it pertains to health and wellness and adequate living space for the entire flock. Providing roomy living quarters, isolating injured birds as soon as possible, and eliminating harsh artificial lighting (in both the brooder and in the coop when augmenting light during the winter) are all steps toward reducing the stress that leads to cannibalism. The same precautions outlined in "Keeping the Peace" on page 183 are also used to limit cannibalistic behaviors among flock mates.

It's critically important that you take cannibalism very seriously. Chickens are ruthless in maintaining a strong social structure. After all, any weak, ill, or injured birds will slow down the rest, making the entire flock a target for predation or the spread of disease. To the outside observer, cannibalistic chicken behaviors can seem pretty callous, but these behaviors are their means for survival. It is only

Many popular idioms can be directly tied to flock life and are rooted in very real chicken behaviors. For example, if chickens have spent a large amount of time locked indoors, they can begin to feel *cooped up* and take it out on their flock mates in the form of cannibalism. In humans this feeling is akin to cabin fever, something most of us have felt at one time or another.

when conditions are inadequate that cannibalism can turn deadly. If you see or suspect cannibalistic behaviors in your flock, take action immediately. Investigate the situation and determine the source of cannibalism. Is a hen injured or ill and now being targeted? Are the chickens bored? Is the coop too small? Are newcomers struggling to integrate? Remedy the situation by determining the cause.

Integrating New Birds

There comes a time in every chicken-keeping career when the flock grows thin from losses or old hens stop laying. When this happens, it's natural to want to add new birds to the flock. Before you bring new chicks home, it's important to know a few things. Most importantly, adding new birds to an existing flock is not without its risks. Moving birds to new flocks can lead to the spread of disease, brutal disputes, and sometimes death. Even if all of the birds are healthy, integrating groups of chickens into the same flock can be stressful and disruptive to the social structure.

As you've just learned, the flock's hierarchy is vital for keeping the peace. When each member knows her place in the social order of the group, the flock is calm, content, and (usually) relatively peaceful. Adding new birds disrupts this social harmony and causes stress as the birds attempt to reestablish themselves in the hierarchy. This stress, like many other kinds of stress, can manifest in varying degrees of cannibalism.

In an ideal world, new birds would never be added to an existing flock. For many of us (this author included), that is an unrealistic expectation. Whether your hens mature and egg production wanes, you cull your breeding flock and raise new chicks, or some are injured or lost, most chicken keepers undoubtedly want to add new birds over the years. Few of us have the luxury to house and care for separate flocks of chickens, so integrating is the only answer.

When you decide to add new birds to your existing flock, take a few precautions first.

Match Size to Established Flock. Add new birds only when they are fully feathered and the same physical size as the members in the established flock. If you are rearing new chicks to add to your flock, this step is critical. Chicks raised *with* a broody hen in the flock's environment stand the best chance at survival growing up with the flock. Never put young chicks in with a full-grown flock of hens *unless* they were raised by one of the hens from hatching. Unknown chicks are often viewed as a threat and killed by others in the flock.

Consider the Breed. Do not integrate very submissive birds or "target" birds into a flock of hens of an aggressive breed. For instance, breeds with topknots or feathers around their faces that may obstruct their view, such as the Polish, are naturally submissive. Their limited vision makes them rather skittish and flighty, too. Integrating birds like this into a flock of very aggressive hens could result in long-term scuffles and miserable birds.

Add at Least Two. When adding new chickens, add at least two, preferably more, to the established flock at a time. This reduces the probability of a single bird becoming a target for cannibalism.

Boost Immunity. Boost your flock's total immunity to prepare them for the transition by adding apple cider vinegar to their drinking water, offering a mash mixed with yogurt for healthy probiotics, or, give raw garlic in mash form or through their drinking water. These natural immune boosters are inexpensive and easy to make. (See chapter 7 for additional immune booster ideas.)

Add a Rooster. Consider adding a rooster. A mature cock will quickly establish himself as the alpha of the flock, in turn reducing the fighting among hens for the top spot.

Reduce Stressors. Reduce any other stressors before integrating birds. Stressors such as poultry shows, long hours or days of travel, and the strain of transportation can leave the flock irritated and distressed, having to reestablish the pecking order all over again.

When you're ready, here are five tricks for integrating two flocks:

1. First, quarantine any new adult birds from the established flock for three to four weeks before making the integration. This allows time to watch for signs of illness and disease. It also provides a chance to treat for any health-related problems in the new birds (ideally, you bring home only healthy chickens). If the birds you are integrating are youngsters that you have raised yourself and have spent time on the same soil as the main flock, they are exempt from this isolation period.
2. Let the two flocks hear and see each other during the quarantine period by setting up housing accordingly. Of course, both flocks must have safe living quarters, roosts for sleeping, nest boxes, and food and water.

High, Higher, Highest

The higher a bird perches, the higher she is in the flock's pecking order. If your coop sports a ladder as a roost, with different levels of perch space available, the alpha bird(s) will gravitate toward the top when it comes time to roost. If there's a rooster in your flock, he'll almost always be at the very top. A flock's hierarchy can change over time as birds become sick, weak, die, age, go broody, or if they've been removed from the flock for a time. Take a peek at your flock while they roost at night to get a sense of where each individual bird falls in the group's pecking order. You may be surprised.

3. After the quarantine period, give everyone some free-range time together to get acquainted. This gives both flocks the opportunity to meet-and-greet but also allows everyone the space to take cover and hide if need be.
4. Gradually move the flocks closer. Move the new birds into the main flock's housing with a barrier. A dog crate, kennel, or partition fenced off is ideal for this part of the transition.
5. When you feel they are ready, place the new birds on the roost with the main flock after dark. This will reduce initial fighting and allow the birds to "wake up" with each other the next morning. Keep everyone together in the enclosure for the first few days.

Remember, no matter how smoothly or gracefully two flocks of chickens are integrated, there will always be some fighting and show of aggression as the birds work out their new pecking order. The fighting may be as little as a few pecks here and there, or it could result in severe bullying. Always watch newly integrated flocks very closely during the first few weeks. Look for signs of injury, such as blood, loss of feathers, or voluntary isolation. Treat accordingly. Also, monitor all of the birds to be sure they have access to food and water and that no one is getting chased away in a show of dominance. As long as everyone is eating and drinking, don't be alarmed if egg laying slows down in the first few weeks. It will pick back up when they've all adjusted to their new flock mates.

On rare occasions, some birds will never get along. It's rather uncommon, but it does happen. If you find your flock in constant turmoil with endless fighting, cannibalism, and/or starving birds, consider how important it is to you that you keep all of these birds together. Instead, consider culling (removing) one bird or two from the flock and see how the dynamic changes. In my experience, removing the bully allows the flock to reorient its hierarchy. When the bully is returned, bickering may end altogether. If it doesn't, consider making the choice to permanently remove one or more birds from your flock.

Broodiness

Broodiness is the maternal instinct of a hen to sit on a nest (or clutch) of eggs until they hatch. Just as she is going broody, a hen will lay up to ten eggs, and then get comfortable—literally and figuratively.

Chicken Chatter
We know that when a rooster is part of a flock of chickens, he will almost certainly take the dominant spot as the alpha of the social hierarchy. He settles disputes and commands respect in exchange for his duties and services. In the chicken world, he is the one who often *rules the roost.* In the human world, to *rule the roost* is to be the boss or to be in charge, most often of the home.

She'll find a cozy nest, turn over the bedding, fluff up her bedding, and get cozy. Her goal is to sit on her clutch until they hatch, about three weeks later.

A broody hen is often protective of her nest and will issue warning growls or vocalizations to intruders.

While most of our favorite breeds of domestic laying hens have long since had the brooding instinct bred out of them, some breeds still have a tendency to go broody. How often this happens depends on temperament, too. Some individuals, regardless of their breed, are more prone to broodiness than others. Usually in conjunction with the shorter days of the fall, a hen may experience a surge of the hormone prolactin, released from her pituitary gland, that signals her body to stop laying for a time. The proud mama-to-be focuses all of her attention on her new task and won't be deterred.

Broody hens tend to get a bad rap. For those of us concerned with egg production, finding a broody hen can be frustrating. To boot, they can be a bit ornery, refusing to eat, drink, or lay eggs, and they may even give you a "fowl" growl when you approach the nest. But brooding is only natural. How else would we hope that chickens reproduce?

How do you know if your hen has gone broody? Here are some things to watch for:

- Does she sit on the same nest, day after day, even after you've removed the eggs from under her?
- Does she puff up or ruffle her feathers when you come near?
- Does she offer a defensive, low growl or even a hiss when she's been disturbed?
- Has her comb and wattle gone pale (from not leaving the nest to eat or drink)?
- Does she seem undeterred and focused on the task at hand? (This one is subjective, of course, but it counts.)

Set a Spell

With a warm nest beneath her and hormones surging, it can sometimes be an impossible task to break a hen of a broody spell. Much depends on her breed, individual disposition, and how long she's been sitting. As a rule, the earlier you catch a hen entering a broody spell, the better your chances are for breaking it. Don't be discouraged if your hen consistently thwarts your efforts; some hens are just stubborn. Some otherwise docile hens become aggressive or defensive of their nest. Be careful and be respectful if you try to break a hen's broody spell.

If you suspect that one of your girls is going broody and want to keep her from setting, act fast. After a few days of brooding, the following tips are less likely to work.

Ice. The conditions that a broody hen strives for in her laying and hatching environment are *dry*, *warm*, and *dark*. By removing some of these elements, you may have a chance at throwing her off her game. One old-fashioned method is to put a few ice cubes under the hen (after removing any eggs, of course). With this method, you'll inevitably end up with a soggy nest box but maybe one less broody hen.

Water. A similar method removes the hen from the coop and keeps the messes outside. In an attempt to cool off the hen's bottom, you may try gently dunking her underside in a basin of cool water. Try this method several times a day, in conjunction with temporary new housing (see below). Avoid this method in very cold weather conditions.

Wire. A less-messy and lower-maintenance approach is to move your broody hen to cool, drafty, short-term housing. Combating her desire to be warm and dark, remove your broody hen and house her temporarily in a wire-bottom poultry cage or pen with good airflow beneath the cage. (This is the one and only time that wire-bottom pens and drafty housing will be recommended.) Provide food and water, of course, but do not give her a nest box or nesting material. Try this method for only a short amount of time—a few days to a week.

Broody hens aren't all bad; in fact, many backyard chicken keepers find immense joy at hatching their own eggs with one of their hens. Children especially love to watch chicks hatch, learn to forage, and grow up with their mother hen—it can be a valuable and beautiful life experience. Letting your broody hen hatch her own eggs is also a great way to add new birds to your flock safely. Just be prepared to rehome any roosters if your city forbids them.

A great way to add new birds to a flock is to let your broody hen hatch her own eggs.

Rooster Rules

Most municipalities do not allow roosters within city limits and with good reason: They're really noisy. And though our collective pop culture association with roosters is early-morning crowing, they don't reserve their boisterous vocalizations for daybreak alone. Roosters crow all day long. Roosters are always on the alert for danger, so they'll crow if they hear or see anything loud or out of the ordinary—that could be kids playing, a car alarm, or a door slamming. The rooster's crow is also his territorial calling card. He's letting other males in the area know that this is his turf and his flock.

If your city *does* allow roosters, you may want to consider reading on. Roosters can and do play an important role in the social dynamic of chickens. Keeping a rooster with a flock of laying hens can be beneficial in many ways. Here are a few to consider.

The Pros

- Roosters protect the flock from predators. A good rooster—meaning one who does his job—will put himself between his hens and danger. If a predator attacks, he will fight the intruder, to the death if necessary. If he's not able to fight, he will do his best to buy some time to allow the hens to get away. He's always on the alert with one eye to the sky and anywhere else a killer may be lurking.
- A rooster means fertilized eggs. If you dream of hatching your own chicks, there's no better place to start than with your own fertile eggs. Roosters "service" (that is, copulate with) their hens throughout the day, so if there is a young, healthy rooster among your hens, you can be fairly certain they are laying fertilized eggs.
- Keeping a rooster in your flock will eliminate many social disputes. As long as he is mature, most roosters move immediately into the alpha position within the hierarchy. Hens will still squabble to establish their individual status in the rest of the flock's pecking order, but fights will be significantly reduced. If scuffles between the ladies continue, the rooster will step in and put a stop to it.
- A rooster ensures that everyone's needs are met. When he finds something tasty to eat, he'll croon to his ladies and let them have first dibs.

The Cons

- For the same reason that roosters make good protectors, they can also be poor additions to family flocks. They are aggressive by nature. We all know someone who grew up on or near a farm and had a poor, if not traumatic, experience with a nasty rooster. For their size, they are incredibly strong and their spurs long and sharp. They move quickly and fight dirty, too. They'll wait until your back is turned to strike. That said, there are exceptions to every rule, of course. Some roosters can and do make excellent family pets.
- Roosters do indeed service hens regularly. Some males have only one thing on their mind, to the detriment of the flock. These roosters can inflict serious damage on a small flock of hens. Torn combs and pulled-out neck feathers (where the rooster grips onto the hen) and missing saddle feathers and exposed skin from the rooster's feet are some of the injuries an overly mated hen may experience. The most dominant hen in the flock

is usually the rooster's favorite, and he'll mate with her more frequently than the rest.

- Roosters are indeed noisy. They'll crow throughout the day, and loudly. If you live very close to a neighbor, or if the coop is near a window or close to your own home, consider how comfortable you are with the noise before adding a rooster to your flock.

Of course, there's absolutely no harm in keeping a ladies-only henhouse. Hens will lay eggs whether or not there is a rooster around since a hen's laying cycle is determined by her own hormones, not the presence of a male. Of course, you'll need a rooster if you want fertilized eggs.

Chores

The rumors are true: An entire flock of backyard chickens is easier to care for than a single house cat. Aside from coop setup, their daily feeding and watering, basic maintenance, and need for human attention are very minimal.

Protection from predators, fertilized eggs, and social harmony are a few good reasons to keep a rooster around.

A Chicken for Every Season

One of the reasons that chickens are so easy to care for is that the major (and messiest) chores happen only once or twice a year. The daily care for a chicken truly takes less time and stress than scooping out a litter box. Below are some of the daily, weekly, monthly, and yearly chores to expect when caring for your own flock.

Daily Chores

I like to think of my daily chicken chores as a dance of giving and receiving. I bring fresh feed out to my flock of ladies, and I return to the house with warm eggs. I even use the same stainless steel pail for both tasks, so it feels a bit like a good trade.

Daily egg collecting is best practice, since it keeps eggs from being stepped on, becoming dirty, rotting in the heat, or freezing and cracking in the cold. Keep an egg basket near the door to remind yourself to collect eggs during rounds, or use the same feed pail and fill it up after you feed your flock.

A full feeder can sit for a day or two at a time, but it's best practice to change water daily, especially if birds are prone to perching on the font and soiling it. Fresh water is critical at all times of the year. In the winter, keep two fonts to switch out when one becomes frozen.

Another daily chore is a visual inspection of your birds. Glance around the coop or run and scan the chickens for any visible problems. Do all of the birds look healthy? Are they engaging in normal chicken behaviors, such as pecking, scratching, dust-bathing, and preening? If not, they'll require a closer look to determine what, if anything, is wrong.

If you're in a hurry, not to worry if these chores are not completed in the morning. They can easily be done at any time of day.

Weekly Chores

The beautiful thing about caring for chickens is that they are not picky. If you don't get to your weekly chores on time, they won't hold a grudge (unlike some felines I know). Chickens are forgiving that way.

Most weekly chores revolve around coop upkeep and basic maintenance. On a weekly basis, you may find that you need to replace or add new bedding in the nest boxes or in the coop. (I have a hen that never fails to step on and crush an egg here and there, making a goopy, sticky mess.) Remove crushed eggs, shells and all, plus droppings and other debris from the nest boxes, and then add additional bedding. Most of the bedding will still be clean, so don't be wasteful; just remove what is soiled and add enough to have a soft place for the ladies to lay their eggs.

If you have sand in your run or coop for drainage, now is a good time to rake out droppings and turn it over. Clean out droppings boards from under the roosts. While you're in the run, do a perimeter check of the inside and outside of the entire structure. Look for signs of predators, such as digging and holes, claw marks, or weak spots in the coop that a predator could get through.

Once a week, give the water font a good scrub with warm soap and water. Be sure to rinse it out thoroughly. Adding apple cider vinegar to the birds' water is also a nice way to keep algae down and give the flock a little supplemental immunity boost at the same time. While you clean out the water font, give your birds a fun treat to distract them and keep them occupied.

Collecting eggs is one of the most exciting daily chores, for children and adults alike.

Monthly Chores

Now's the time to do the chores noted during your weekly inspections. If the coop is in need of repairs, set aside a weekend to take care of them, especially if repairs are putting your birds in danger from predators. Put those tasks at the very top of your to-do list.

Plan to swing by your local feed and seed store and stock up on layer ration. Once the fall rolls around, it's wise to have at least two months of feed purchased and stored at home for the flock. Chickens tend to eat much more in the colder months, going

Checklist for Flock Care

The following checklist provides a handy reference to remind you of the most important weekly, monthly, and yearly chores required when caring for your own flock. Make a copy and hang it where you can see it regularly—on the refrigerator, in the coop, or wherever you will be sure to walk by it daily.

Daily Care Checklist

- ☐ Give fresh feed and water.
- ☐ Collect eggs.
- ☐ Do a visual "once over" of each bird, checking for obvious injury, blood, or changes in temperament.
- ☐ Do a visual "once over" of the coop and housing, inspecting for any predator attack attempts.

Weekly Care Checklist

- ☐ Check supplements (oyster shell and grit) and provide more, if necessary.
- ☐ Offer healthy treats and kitchen scraps.
- ☐ Replace soiled pine shavings and/or add more bedding, if necessary.
- ☐ Rake or turn over sand in the run.
- ☐ Turn over litter in the coop.
- ☐ Scrub out water fonts.
- ☐ If using a droppings board, monitor buildup and clean out, if necessary.

Monthly Care Checklist

- ☐ Purchase feed, oyster shells, grit, and other sundries, as needed.
- ☐ Check supplements (oyster shell and grit) and provide more, if necessary.
- ☐ If using the Deep Litter Method (see chapter 6), add another 1 to 2 in. (2.5 to 5 cm) of bedding.
- ☐ Check the coop for damage and make repairs, as needed.
- ☐ Hold each bird and do a physical inspection for injury or illness.

Yearly Care Checklist

- ☐ Replace bedding and completely clean out the coop.
- ☐ Order and replace broken equipment, as needed.
- ☐ Order and raise new chicks, as needed.

through feed faster than in the summer. If you live in a snowy region, winter storms, weather-related emergencies, and other obstacles could make it difficult to obtain feed when it's needed most. Better to be safe than sorry, so stock up.

Next, check levels of grit and oyster shell. Remember, your chickens will take their supplements as their needs dictate. Your job is to make sure they are always available and to refill when rations get low. Over time you'll come to know how much your flock needs, depending on the number of birds you keep, how often they free-range, how strong a layer they are, and other factors.

Finally, handle each bird to inspect for signs of injury or illness. Check for broken wings, broken toenails, or injured feet and to see that the bird's eyes and beak are clear of discharge. Check for external parasites, signs of internal parasites, and check the vents and under the wings for evidence of mites. Once practiced, the whole process will take no more than a minute or two per bird and could give you a leg up on catching problems early.

Yearly Chores

To keep a chicken coop is to keep a big box of mess in your backyard. Therefore, there's no avoiding the biggest, often most dreaded chicken chore of all: the coop clean.

Your management practices will determine just how big and how messy your yearly coop cleaning will be. However you do it, keep a few basic pieces of equipment on hand: a heavy-duty pair of disposable gloves (rubber kitchen gloves work well), face mask for dealing with the inevitable chicken dust, shovel, rake, scrub brushes, and a light nonchlorine bleach-water solution (as an alternative, you could use a 50/50 vinegar/water mix as well). You'll also likely benefit from access to a hose and a bucket or two, depending on your disposal methods.

To do a deep clean of the coop, first remove all of your birds from the area you intend to clean. Not only will they get in your way and thwart your cleaning efforts, they'll likely be put off by such an intrusion. Remove all roosts, nest boxes, waterers, feeders, heat lamps, and any other objects from the coop. Then remove all bedding and clean the structure. This bedding can make a wonderful addition to a compost pile, so don't throw it away. Once the coop is dry and aired out, replace the bedding and mix in several handfuls of diatomaceous earth. Return all of the objects to the coop and allow your flock to mosey on back at their leisure.

Rehoming

Until this page, the main focus of this book has been on preparation for becoming a steward of chickens. It was written to encourage you to thoroughly research, reflect, and then decide whether chickens fit into your life, your routine, your family's dynamic, and your town or neighborhood. As with any pet or animal in your charge, taking on chickens requires a commitment of time, money, and education. Once you bring home a box of peeping chicks, those creatures become your responsibility, 100 percent.

How to Get Out of Keeping Chickens

On occasion, it happens that, even with all the research and preparation in the world, some chickens and humans aren't the right fit. Sometimes, neighborhood associations change their regulations. Or landlords change their minds. Sometimes, an ongoing neighborly dispute ends up in too many calls to animal control officers, forcing you to give up your birds. Maybe you jumped the gun and started a flock before your city legalized chicken keeping, and your flock was reported. Of course, some people find that the responsibilities of keeping chickens, easy as they are, are too specific to fit their lifestyle. Whatever the reason may be, rehoming a flock of chickens requires special consideration and some planning.

Your rehoming options greatly depend on your location and how "chicken-friendly" your town, suburb, and surrounding area may be. Here are some of the most popular choices.

Off to the Farm. If you're very lucky, you may find a nearby farm that will take in your flock. Ideally, the farm is traditional in that it provides its chickens with ample outdoor forage space, green pasture, and a safe barn or coop. Locating the right farm might take a bit of legwork. Ask at feed and seed stores, farm supply stores, and even local grocers, because they may have recommendations. Call ahead before showing up. Never drop off chickens without clearing it with the owner of the property first.

A New Backyard Home. As chicken keeping gains in popularity, prospective keepers may be on the lookout for a small flock to call their own. An ideal owner is someone who already has a coop established, has his or her ducks in a row (necessary permits, permission from homeowner's associations, and so on) and prefers to start with an adult

Chicken Sitters

Most new or prospective chicken keepers shudder at the thought of being tied down to their home and unable to travel because of responsibilities to their flock. For some potential chicken keepers on the fence about getting chickens, this can be a deal breaker. But what if I told you that you could have your eggs and eat them, too? Many traditional pet sitters are taking on poultry "clients" with excitement and a growing body of expertise, allowing you and your family to travel near and far and still have the peace of mind that your family flock is being well cared for. Here's how to go about finding the right sitter:

1. Join a local chicken-keeping club or get connected with other chicken keepers in your area. They may have monthly meetings or a Facebook page where you can connect. Many will trade chicken-sitting services or watch your flock for a fee.
2. Hang a Wanted ad at local feed and seed stores, farm supply stores, local universities or colleges, or veterinarian offices. Be specific about what you are looking for and include contact information. (Remember to always ask the owner of the establishment before posting anything.)
3. Post a Wanted ad through Craigslist (craigslist.org) or another online site. Remember to categorize it appropriately, using either the "Community" or "Pets" categories.
4. Ask your current pet sitter or other pet sitters in the area if they would take on poultry. If they're not already familiar with chickens, offer to teach them the basics of poultry care. Remind them that it's super-easy.
5. Ask local friends and family if they'd like to be a farmer for a day—and let them keep the eggs they collect to sweeten the deal.

flock of hens rather than raising chicks. To find these folks, network using online resources such as Facebook and/or tap into local poultry clubs. Put the word out to friends and family in the area. Post on Facebook, either on your own Timeline or in local homesteading/poultry groups. Ask them to share your post on their Timeline with all of their friends. Word-of-mouth might be your best bet for finding a suitable home for your hens.

While ideal, it would take some networking, plus a bit of luck, to rehome a flock of chickens to a humane, traditional farm.

Another option is to post For Sale ads. For print ads, head over to feed and seed stores, farm supply stores, and veterinarian offices. For virtual ads, Craigslist is likely your best resource. Post under the "Community," "Pets," and/or "Farm" categories for the most views.

Shelters. Surrendering your flock to the local county animal shelter should be an absolute last resort, used only in an emergency or urgent situation. Many animal shelters are not equipped to properly house, feed, or care for chickens. Some rural shelters may pick up the occasional stray bird but rarely are they outfitted to handle their care long term. If you absolutely *must* surrender your flock to animal control, provide as much information about them as you can to the animal control officer. Most importantly, provide their ages and report any vaccinations the birds have had. If possible, provide a written record. Also, give the officer(s) a history of the chickens' origin, such as the breeder, hatchery, or farm where they were hatched. Once you have dropped off your flock with the county shelter, continue your campaign to find them a good home by posting photos on Facebook and ads on Craigslist and other sites, including your local newspaper and farm shops. Do your best to give them a good second chance.

Rehoming a flock is rarely easy and not without its stressors. If you're under pressure from an organization or the city to remove the birds from your property, do your best to buy a little extra time. Explain that chickens have special needs and are unlike traditional pets, requiring just the right home and a unique living situation.

Finally, one option that should *not* be considered when getting rid of your chickens is releasing your flock to fend for themselves. Releasing chickens creates feral populations that can become significant city problems. More likely than that, however, is that your birds become prey to one or more of the many predators that lurk nearby. Releasing a flock of hens "into the wild" will more than likely spell their certain death and is irresponsible at best.

Chapter 12

The Perfect Food

The domestication of the chicken—something that is thought to have taken place about 8,000 years ago in Southeast Asia—made it easy for humans to acquire chicken eggs without disrupting their reproductive cycle. With a clearer understanding of the seasons and the female chicken's reproductive patterns, humans throughout time and across cultures have been able to maximize egg yields and make the egg the culinary staple we know today.

This chapter covers everything you would need or want to know about the egg, from basic anatomy and its nutritional breakdown to proper egg care and storage. Because not every egg is perfect but all eggs are unique, later pages are devoted to the odd and unusual eggs you will encounter as a backyard chicken keeper. No need to be alarmed; imperfection is part of the beauty of life, with chickens and otherwise.

The Anatomy of an Egg

The obvious reason for keeping chickens is for their high-quality contribution to the breakfast table. There's nothing better than freshly gathered eggs, which are certainly better tasting than the mass-produced eggs found at the grocery store. But have you ever wondered exactly what's inside that eggshell? Well, wonder no more. Below is a guide to understanding the anatomy of an egg.

An egg is composed of several components: the yolk, the albumen, and the shell. Let's look at how each of those parts functions.

The Yolk

The yolk is the thick, round, yellow-orange part of the inside of a chicken egg. It consists of a generous portion of protein and all of the fat of a chicken egg. The yolk's purpose is to nourish the developing embryo through incubation and to provide the day-old chick with sustenance for up to three days following hatching. With the remainder of the yolk sac absorbed, a newly hatched chick will remain nourished, waiting for the rest of the clutch to hatch.

Greens Are the Key

The yolk's color deepens in hue depending on what the hen has eaten. The addition of greens to a hen's diet contributes to a darker, sometimes orange-colored yolk—in contrast to the pale yellow yolks from commercial farm eggs.

The Albumen

More commonly called the egg white, the albumen is the clear liquid that surrounds the yolk within the egg. Transparent when in its liquid form and white when cooked or heated, the albumen is part water and part protein—about 40 different proteins in fact. Composed of four separate compartments—the outer thin, the outer thick, the inner thin, and the inner thick—the egg white's two main purposes are to repel bacteria and cushion the yolk.

Within the albumen are also two cords called chalazae (chalaza singularly). These two cords, one connected to the "top" of the yolk and one connected to the "bottom," suspend the yolk within the egg, thereby keeping it stationary and properly orienting the fragile insides for hatching. You're probably familiar with the chalazae: When an egg is cracked open, the chalazae snap back toward the yolk and appear as tiny, white knots on the yolk's surface.

The Shell

The shell is the hard, calciferous outer structure of an egg. The shell of an egg consists of three layers: (1) the inner layer connects the albumen to the shell; (2) the spongy layer has pores that allow the flow of oxygen and helps release moisture and carbon dioxide from the shell's interior; and (3) the cuticle layer, located on the shell's outside, seals the egg's pores, preventing bacteria from accessing the inside. This last layer, also called the bloom, is water soluble and is washed away when the egg comes into contact with water, removing the protective layer and exposing the egg's interior to bacteria.

The Perfect Protein

Eggs are among the best sources of protein in the diet, containing all of the essential amino acids (the building blocks of muscle). Learn about the health benefits and the nutritional facts below.

The Yolk. A pastured chicken's egg yolk is a rich, natural source of vitamins A, D, E, and K, healthy fats, "good" cholesterol, and lecithin, a natural emulsifier that helps break up fat and ease digestion. Egg yolks are also a great source of choline, an essential nutrient that contributes to healthy brain development and liver function.

Tales from the Coop

When my teenage sister, Bella, came to visit us in North Carolina one summer, she refused to eat our eggs. Since she was a lifelong devotee of the runny scrambled-egg-with-cheese, I attempted to bribe and persuade her with all manner of tricks. My attempts were to no avail. The thought that my dear, sweet sister would miss out on the best eggs she had ever tasted made me sad—and just a touch offended.

What she didn't know at the time was that I had also been reluctant to eat our first eggs many years before. I now consider this apprehension (both mine and hers) a remnant of our total lack of awareness about the source of our food. In a small way, the fact that one of our hens had actually laid an egg (it came out from under her tail!) was a tad off-putting. It was a bit of a culture shock to truly see where and how an egg was produced. She and I were so accustomed to being distanced from our food that being up close and personal made us uncomfortable.

These days, I'm grateful for the perspective that I have (and keep) on the process of growing food and food production. Creating food means creating life, and that is not always a pretty business, nor is it perfect, like those supermarket eggs I had grown up eating. But these days, I wouldn't trade my knowledge for anything. A new experience, a tiny shift in perspective, and a tasty egg can change your life for the better.

Chicken Lore

There is no nutritional difference among white, brown, or colored (blue, green, olive) eggs. The hen's diet, quality of food, and accessibility to forage (for grasses, greens, and bugs) determine the quality of her eggs.

The Albumen. A tad bit higher in protein than the yolk and full of vitamins B_2, B_6, B_{12}, selenium, and trace minerals, the egg white has long been a favorite of dieters for its lower-fat content, low calories, and exceptional protein. Of the two main parts of the chicken egg, most people with sensitivities or egg allergies react to the albumen, not the yolk.

Better Together. The yolk and the albumen are nutritious and rich foods individually, but together, a whole egg provides a compact source of protein that has no equal. In fact, the egg white and egg yolk must be combined in order to reap some of the benefits that the egg has to offer. For example, the selenium found in egg whites works with the vitamin E found in egg yolks to help prevent the breakdown of body tissues through its antioxidant qualities. Lutein and zeaxanthin, carotenoids that support healthy eye function, may help prevent macular degeneration that could lead to blindness in aging adults. A whole, single egg also contains all nine of the essential amino acids that humans need for healthy immune function and to combat the absorption of excess fat. Considered the building blocks of a healthy body, amino acids contribute to muscle growth and the daily repair of tissues, organs, and muscles. The catch with essential amino acids is that they must be obtained from a food source since the human body cannot manufacture the acids on its own—and luckily for us, the egg happens to be the perfect source.

Omega-3s. Another element that is essential to good human health is omega-3 fatty acids. Like amino acids, we must source these fatty acids from the foods we eat, since the body can't synthesize them. Eggs from hens with a varied diet, but especially eggs from pastured hens, will naturally be higher in omega-3s. Some large-scale egg producers who tout and sell eggs with "boosted" omega-3 fatty acids have fed their hens supplements—such as flaxseed, kelp, fish oils, or marine algae—to enhance the eggs' omega-3 content. Your backyard flock, if allowed to free-range on grass, will give you eggs rich in these fatty acids.

The egg's yolk and white work together to make a nutritional powerhouse.

The Egg's Bad Rap

Once a mainstay of the human diet, the egg is only recently returning to its rightful place on the plates of health-conscious eaters. For many years, health professionals told patients to limit or entirely avoid eggs due to their cholesterol content, mistakenly believing that dietary cholesterol was the primary cause of high blood cholesterol levels. The egg was quickly demonized as the harbinger of ill health and avoided by many. But, in 2000, the

American Heart Association amended its guidelines, once again recommending eggs for a healthy diet. Why? The research had spoken: Studies found that high blood cholesterol levels were in fact more influenced by the intake of saturated fats, not dietary cholesterol. In fact, the lutein found in egg yolks had been found to protect *against* some early signs of heart disease. In moderation, it was now understood that the regular consumption of eggs could, in fact, be good for you.

And, as you already know, healthy chickens make healthier eggs. Those sourced from pastured hens offer the highest nutritional benefit you can get from a chicken egg. While the egg is still vying for its role in the modern American diet, you can help fight the myth by educating others and, of course, sharing, selling, and eating this incredible food.

Chicken Lore

The egg has long been a symbol of birth, regrowth, and fertility. Some ancient cultures viewed the yolk as a symbolic sun; others used eggs for divination, believing they could see into the future by reading the cracks after boiling or by tossing them whole. Creation myths out of Europe, Asia, and Africa speak of a cosmic egg, from which the heavens and Earth were hatched. In one of those myths, the egg hatches seeds of possibility, springing forth the start of everything.

Today, eggs are undoubtedly a harbinger of spring. As the days grow longer, hens resume laying, and eggs become plentiful again, reminding us that winter does come to an end and that fruitful days lie ahead.

Keeping Eggs Clean

No one likes a dirty egg. Fortunately, if you're keeping a clean coop, dirty eggs will happen only once in a blue moon. As with any other health-related chicken issue, prevention is worth a pound (of eggs) of cure. Practicing good sanitation practices like those listed below will keep the majority of your eggs in tip-top, sparkly, clean shape.

- Clean nest boxes regularly by removing visible droppings and broken eggs (including shells) and replacing bedding regularly.
- Keep flooring in the coop, run, and any outdoor enclosure tidy so chicken feet remain clean (ish). Take special care during continuous bouts of rain, heavy storms, inches of melting snow, or any other soggy conditions that could lead to mud puddles and muddy feet.
- Keep nest boxes for laying and roosts for sleeping. If you see old hens or young pullets (the two most likely culprits) using nest boxes for sleeping, examine your coop setup, since they may be getting confused. Remember, chickens like to roost high and nest low, so you may have to adjust your coop's layout, move roosts around, or add new boxes if your eggs keep getting dirty. "Retraining" layers is a nuisance, but it's not impossible.

Washing an egg should be a measure of last resort reserved for only the dirtiest of eggs. Contrary to what our germophobic, hand-sanitizer–happy culture prefers, more often than not, washing poses more of a health risk than not washing—at least when it comes to backyard eggs. As discussed earlier in this chapter, each egg is laid with a protective layer called a bloom. This layer keeps bacteria from penetrating the shell and wreaking havoc inside. When an egg is washed, the bloom is destroyed, leaving the egg vulnerable to contaminants. What's more, a wet eggshell is more susceptible to drawing bacteria than a dry one.

The Egg's Role in Cooking and Baking

Historically, the egg has dutifully worked in the kitchen as a binding and thickening agent, an emulsifier, a glaze, a flavor enhancer, and so much more. It's helpful to understand the science around how the egg works its magic.

Many of the egg's heroic feats in the kitchen can be attributed to the proteins in the egg and how these proteins react to different elements, temperatures, and applications. For example, both the proteins and the moisture content (egg whites are over 80 percent water and yolks over 50 percent) contribute to leavening. When heated, the moisture in an egg is converted to steam, creating air pockets in the final product that give the food a boost in texture, while the yolk enhances the food's richness and flavor (think cookies, brownies, and cakes). Of course, the high water content in an egg also lends rich moisture to foods, as well. The proteins contribute to leavening by a different process. When rapidly beaten, the proteins unwind into a flexible film that encloses air bubbles, resulting in a light, fluffy texture that holds its shape (think meringues and soufflés).

The egg's proteins don't stop working there. When an egg is heated, for instance, the proteins congeal and work as a binding agent to hold other elements together and build structure—think of the way a meatloaf holds its shape.

That's not all. Lecithin, the fat emulsifier located in the egg's yolk, is also a hard-working element when used in cooking. The egg's lecithin content helps to break apart fat and suspend it together. This emulsification is critical to getting the right velvety texture for sauces like hollandaise and condiments like mayonnaise.

Finally, egg whites may be put to work clarifying soups into a crystal-clear broth. When whipped and heated, the albumen binds to and traps the tiny particles in the soup. Then, once cooled, the egg whites (and the particles) stay in place and are easily removed. Finally, a beaten egg can be brushed onto breads, pastries, and other foods before baking, creating a coating that leaves the final product with a mouthwatering sheen.

Realistically, though, occasionally an egg does get soiled. The most common culprit is a hen with dirty feet entering a nest box and stepping onto eggs laid earlier in the day. Another common way for eggs to get dirty is by coming into contact with a broken egg. An occasional soft-shelled egg plus a heavy-footed lady is a recipe for a gooey, sticky, nest box mess. Don't beat yourself up. It happens. Despite our best efforts, sometimes an egg needs washing.

So, what's the best way to wash an egg when you have to?

First, you'll want to use water that is warmer than the egg itself. Remember, an egg's shell is incredibly porous. Those pores contract when they come into contact with cool water, inviting all manner of microorganisms into the egg. Most people find that it's easiest to rinse the egg under warm, running water (though it's best not to soak it). You can use a little unscented natural soap if you want to—just don't use anything with harsh chemicals, dyes, or fragrances. The latter, especially, will impart an odd scent and possibly an off flavor to the egg. When we see chicken poop near something that will become our breakfast, often the first reaction is to use something more abrasive, like vinegar or bleach, but it's really not necessary. Using water is enough. If soap and water doesn't cut it for you, though, poultry catalogs sell egg-washing solutions designed for farmers and sellers.

Contact with water destroys the invisible protective layer encasing the shell, so washed eggs must always be refrigerated to preserve their freshness.

Once washed, dry the egg thoroughly. A thin coat of vegetable oil on the egg's shell may prolong the shelf life of a washed egg, but it's not a required step. Either way, *always* refrigerate washed eggs.

Really, really dirty eggs are probably best discarded or fed to pets. In fact, scrambled eggs with crushed/dried eggshells are a fun protein- and calcium-packed treat to give back to your chickens. And you know they won't object to a little poop.

Macroflocks

According to United Egg Producers, in 1987 there were roughly 2,500 operations producing eggs for the American public. Today, only about 180 "companies" (notice they're not called farms) with flocks of 75,000 hens supply 95 percent of the country's egg needs. Personally, I'm proud to be part of the 5 percent who either buys from local farms or raises my own hens, thereby opting out of farming on a factory scale—and I bet you are, too.

The safest way to store eggs is in cartons, trays, or ceramic dishes designed specifically to hold them.

Storing Eggs Safely

When all of the edible eggs pass the backyard inspection test or washing, you'll need to find a place to keep them.

For washed eggs, the storage space is simple. They must go in the refrigerator. New or recycled egg cartons, ceramic egg keepers, or an egg box that is built into your refrigerator unit are all smart choices for storing eggs. Personally, my favorites are the 30-egg flat cartons. They stack nicely without breaking eggs, and they're very inexpensive. Traditional 12-egg cartons are also safely stackable and make it easy to hand out a dozen at a time to friends and family.

Unwashed eggs are generally less finicky. You can toss them in the fridge, store them in the pantry, or leave them on the countertop at room temperature. It's true; unwashed eggs don't *need* to be refrigerated. They'll actually keep longer in cooler temperatures. Unwashed eggs stored in the refrigerator at 40°F (4°C) will keep for up to six months or more.

Fresher Longer

Store eggs pointed end down, round (or large) end up. This direction prevents an air bubble from forming and keeps the eggs fresher longer.

Wherever you store your eggs—washed or unwashed—make sure the ambient temperature and humidity levels remain relatively constant. Yo-yoing between hot and cold can cause any egg to spoil. The key, especially, is to avoid condensation (which is really just moisture) forming on the shell. This can happen when eggs are moved from cold to hot storage or if they're stored in a humid laundry room.

Another good rule of thumb is to stash eggs away from strong scents. We know that, washed or not, an egg's porous shell will quickly soak up odors, good, bad, or ugly.

Egg Abnormalities

It's easy to be wooed into the illusion that all eggs are perfectly formed, evenly colored, smooth, and all the same size, but nature thrives on imperfection. Flaws are unique and interesting, and odd eggs are par for the course in the world of backyard chicken keeping. On the whole, egg abnormalities are nothing to be concerned about. They're usually still good for the frying pan and a little laugh. At best, most of the following odd eggs will remind you to provide additional calcium, in the form of oyster shells, to your hens' diets. At worst, an odd egg could indicate reproductive stress or a hereditary anomaly. The following pages explore common backyard egg oddities.

The Fairy Egg

Sometimes called dwarf, wind, or fart eggs, these tiny orbs are about the size of a grape and are often yolkless. Another term for this very small yolkless egg is cock egg. As the name implies, it was once believed that roosters laid these eggs and that if they hatched, they would birth a terrifying, serpentlike creature called a cockatrice or basilisk that could kill with a single look. Naturally,

As a chicken keeper, you'll experience all manner of odd eggs firsthand.

Chicken Lore

As an egg ages, the albumen begins to shrink away from the sides, and the tiny air pocket present in the egg begins to grow. Since an older egg has a larger air pocket, it's rather easy to test the freshness of an egg without even cracking it open.

Fill a large, clear bowl with cold water. Gently place the egg in question into the bowl. An egg that sinks to the bottom of the bowl, lying on its side, is rather fresh. If it sinks to the bottom with one end pointed up, it's still good to eat (the floating end indicates a growing air pocket). If the egg floats at the top of the water, it's bad; toss it out or compost it.

the eggs were destroyed. Even today, there's not much use for them, but they are interesting to find nonetheless.

Fairy eggs represent a young pullet's first attempt at laying and are nothing to be alarmed about. As the pullet's reproductive tract matures, her eggs will become larger and more consistent in the size, color, and shape for her breed.

The Soft or Shell-less Egg

Like fairy eggs, soft or shell-less eggs are usually laid early in a pullet's reproductive maturity and are distinguished by a papery-thin, often malleable shell.

For young laying pullets, these eggs are very common and nothing to worry about. Soft eggs are to be expected from older hens, too.

If you come across a soft egg from a healthy adult, it's likely that she has been stressed in some way. Water shortages, drastic changes in lighting, or even loud noises could throw her off-kilter a bit. In the days and weeks following a soft egg, minimize the disturbances in and around the coop. Big moves, new birds, or a predator attack can each stress a flock significantly and can lead to reproductive stress.

If the flock hasn't experienced any recent stress, a soft egg could indicate a calcium deficiency. However, if you find that a hen consistently lays soft eggs, despite having supplements available and being in general good health, it may be a hereditary trait that is out of your control. Consider culling the bird from the flock and refrain from hatching any chicks from those birds to avoid passing on the trait.

The Rough Egg

Rough spots, often called calcium deposits in the chicken world, can look like coarse bumps, bulbous calluses, a sandpaper texture across the shell, or simply discoloration. Sometimes you may find that a hen lays a shell-less egg just before or after an egg with added calcium deposits.

Any roughness on an eggshell is usually a result of a calcium- or a calcium-to-phosphorous imbalance. Without phosphorus, a hen isn't able to absorb and metabolize the calcium she consumes. Hens receive phosphorus through eating bugs with an exoskeleton, such as beetles, so allow them to free-range to find their own source.

Otherwise, calcium deposits are nothing to worry about once in a while from a hen or two. If the symptoms manifest in many birds at the same time, keep a close eye on your flock. They may be receiving too much or too little calcium, especially if the weather has been extraordinarily hot or humid. Those conditions can

Preventing Breakage

Here are a few tips for keeping your eggs intact:

- Keep nest boxes filled with bedding. About 3 to 4 in. (8 to 10 cm) of litter provides enough cushion for the eggs, but it will need to be replaced and refilled periodically.
- Collect eggs early and often. The less time an egg spends in the nest box, the less likely it is to meet its fate with other eggs or clumsy chicken feet.
- Use an egg basket for collecting. You would be surprised how easy it is to drop even just a few eggs on your way from the coop to the kitchen—and I speak from experience. You'll inevitably become distracted, think of something that needs to be done outside, or in a hurry, place an egg on a tabletop. Eggs do roll. Keep a basket just for collecting eggs, and if you have little ones tending to the chickens, ask them to use it each time, too. A basket will keep the eggs safe until they're inside and put away.
- Store eggs in a safe place, away from pets, kids, and heavy objects. This probably goes without saying. If you refrigerate your eggs, don't place other groceries on top of them. Store eggs in something that was made to hold eggs, such as a carton, egg tray, or countertop egg storage solution, such as an egg skelter.

increase a hen's metabolism and reduce the level of calcium available to "make" suitable shells. As always, provide oyster shells in a separate container from the food in addition to plenty of fresh water (necessary for the proper absorption of calcium) and see if the deposits subside.

Finally, if discolored or gritty shells continue to pop up in the nest box, they may be a result of diseases such as Newcastle or infectious bronchitis. If this is the case, the majority of the flock would demonstrate symptoms, and they would occur with frequency. If the flock shows other indications associated with those diseases, consult a vet immediately.

The Russian Doll Egg

Incredibly rare, these double-shelled eggs occur when a completed egg receives a second albumen and shell after reversing direction in the oviduct. This results in a fully formed egg encasing a smaller egg (the smaller one typically being yolk-less), or two fully formed eggs nested one in the other. It is unknown what causes this phenomenon, but the results are bizarre and shocking.

The Double Yolker

When ovulation happens quickly, sometimes two yolks are released simultaneously and become enclosed in the same shell together. This results in one of the most thrilling breakfast discoveries of all time: the double yolker.

Double yolkers result when ovulation happens quickly. They make a surprising bonus when discovered at breakfast time.

Usually much larger than your hen's normal eggs, double-yolked eggs make great eating but aren't the best for hatching. The trait that results in double- or multiple-yolked eggs is inherited and often occurs in heavy-breed hens. Double yolkers themselves are nothing to be concerned about, generally, but, if you find that an individual bird is prone to laying these big eggs regularly, watch her closely for signs of egg binding or a prolapsed oviduct. Some birds have difficulty passing such bulky eggs, and this could lead to bigger problems. Otherwise, enjoy the nutritional boost of two yolks.

Other Curiosities

Far less eccentric (but still not considered top grade for commercially sold eggs) are other abnormalities such as flat or misshapen eggs, blood spots on the shell, or meat spots in the yolks and egg whites. Occasional misshapen eggs may be the result of a frightened or stressed bird. Blood spots on the shell often happen when a young pullet first starts laying, and it usually washes off easily. Meat spots within the egg can happen whether it is fertilized or not, and either way, the egg is still good to eat.

It's also common to find a long or torpedo-shaped egg once in a while. Other than its interesting form (and its tendency to pop up the top of the egg carton), this is a normal, tasty egg. Folk wisdom suggests these are eggs you'll want to consume and not hatch, since round eggs are said to hatch pullets and long eggs are said to hatch cockerels. Either way you crack it, a long, thin egg isn't the ideal shape for *any* developing embryo, so save those for sunny-side up instead.

The Perfect Hard-Boiled Egg?

The perfectly cooked hard-boiled egg is one with a shell that peels off like paper, with no green ring around the yolk and a soft but fully intact white. In my opinion, this egg is an elusive creature akin to a unicorn. In other words, does the perfect hard-boiled egg really exist? Many chicken-keeping experts and world-renowned chefs advertise their foolproof, fail-safe methods for hard-boiling eggs, but I'm here to tell you, I've tried them all; some work better than others, but none of them is foolproof.

So, I'll be perfectly honest. I don't know the trick to the perfect hard-boiled egg. Since I can't be sure, I won't be sharing a recipe with you that promises your eggs will be perfectly cooked and easy to peel each time. This is partly because everyone has an opinion on how to hard-cook an egg, and you'll inevitably see conflicting advice (*always* add salt vs. *don't ever* add salt). It's enough to make your head spin.

Plus, backyard eggs are anything but uniform in size, shape, and texture. I've found with my own personal flock, some of my girls' eggs are easier to peel than others. For instance, Yoko, one of my Easter Eggers, lays a large egg with a round shape and a shell with medium thickness. Hers are always my go-to eggs for hard-boiling. Sadie, on the other hand, is a petite bird and lays a petite egg; her eggs' shells are thick, with pointy ends, and they are impossible to peel, so I keep hers for scrambling. But that's just me.

Even so, I'd be remiss not to share with you a few hard-boiling tips that some chicken fanciers and home cooks swear by. Take or leave the advice, try a few, combine a few, or just stick with what works for you:

- Add salt to the water or. . .
- Don't add salt to the water.
- Add baking soda to the water.
- Use your oldest eggs for hard-boiling. (This one is true. With less moisture inside, the membrane will have contracted away from the shell, making peeling much easier.)
- Place eggs into cold water and bring to a boil or. . .
- Place eggs directly into boiling water.
- Take the cooked eggs out of boiling water and place into a cold-water bath or. . .
- Take the cooked eggs out of boiling water, rinse with cold water, and place in the freezer for 30 minutes.
- Punch a tiny hole into the hollow end of the egg (where the air bubble is located) before boiling.
- Once boiled but before peeling, crack the egg and roll it between your palms.
- Peel while the egg is still hot or. . .
- Peel once the egg is cooled or. . .
- Peel under running cold water.

Got all that? Good. Best of luck and happy hard-boiling!

Acknowledgments

First and foremost, all my love and appreciation goes to my wonderful, patient and endlessly flexible husband, Ian: He is the man who follows my lead with chickens, and with just about everything else in life. Without you, my life would have a lot less direction and a lot more animals. Thank you for your tireless support, your equally matched enthusiasm, and your unwavering trust in me (especially when I bring home new animals).

I'd also like to thank my friends and family who have made this book (and lifestyle) possible: Lynn McNeill-Morejon, for your steadfast loyalty since the age of 14 and for being my biggest fan; Jessie Crutchfield, for your help in caring for my fluffy little "chick" while this book materialized and for offering wisdom beyond your years.

A big "cluck" of thanks to Jonah, Mavis, and their families for agreeing to appear in this book. I thoroughly enjoyed spending time with each of you and your flocks and am so encouraged to see the next generation of chicken keepers already hard at work spoiling their "ladies."

Of course, I'd be remiss not to thank my editor, Dolores York, for her patience, understanding, and like-minded vision as we tackled this project together. Thank you for your guiding light as I navigated these new waters. Thank you also to Mary Ann Kahn and Amy Deputato for putting up with novice questions, rushed late-night emails, and for helping me to create something of which to be proud.

And finally, extra special, massive thanks go to my *Chickens* magazine editor, Roger Sipe, who, years ago, took a chance on an unknown, unemployed, and eager writer and gave her the opportunity to follow her dreams. Without exaggeration, this book was possible because of you.

Resources

The best place to turn to for information on raising chickens in your area is to seek out other chicken keepers in the community, since they are the ones most familiar with some of the issues you may encounter. For assistance with more specific problems, consult the Internet or contact your county extension office or poultry specialists at universities in your state or province.

Publications

Online Information

Backyard Chickens: www.backyardchickens.com

The largest online chicken-keeping community with members from around the world. Search frequently asked questions or post your own to the site's forum. Or, browse coop designs and search through the gallery for member photos.

My Pet Chicken: www.mypetchicken.com

Browse the "Blog" and "Chicken Help" pages for general care information, or shop for equipment, chicks, and fun chicken paraphernalia. This site also has an easy and convenient breed selection tool.

Magazines

Backyard Poultry: www.backyardpoultrymag.com

A magazine dedicated to more and better small-flock poultry care with articles about poultry care and health, coop building, and more. Online features include "Ask an Expert," a member forum, and webinars.

Chickens magazine: www.hobbyfarms.com/chickens-magazine/

The essential poultry publication from the editors of Hobby Farms *magazine. Informative articles on chicken care and health by experts from around the country, as well as tips on coop design, breed selection, illness and injury prevention, and much more.*

Books

Damerow, Gail. *Storey's Guide to Raising Chickens.* 3rd ed. North Adams, MA: Storey Publishing, LLC, 2010.

———. *The Chicken Health Handbook.* Williamstown, MA: Garden Way Publishing/ Storey Communications, Inc; 1994.

English, Ashley. *Keeping Chickens with Ashley English: All You Need to Know to Care for a Happy, Healthy Flock*. New York, NY: Lark Crafts. 2010.

Ussery, Harvey. *Small Scale Poultry Flock: An All-Natural Approach to Raising Chickens and Other Fowl for Home and Market Growers*. White River Junction, VT: Chelsea Green Publishing, 2011.

Willis, Kimberly, and Rob Ludlow. *Raising Chickens for Dummies*. Hoboken, NJ: Wiley Publishing, Inc., 2009.

Equipment

Murray McMurray Hatchery: www.mcmurrayhatchery.com

A one-stop shop for a large variety of poultry equipment, from small- to large-scale poultry care.

My Pet Chicken: www.mypetchicken.com

A good selection of quality poultry equipment, unique coops, and fun-themed gifts for the chicken enthusiast.

Hatcheries

Ideal Poultry: www.idealpoultry.com

Family owned and operated hatchery based in Cameron, Texas.

Meyer Hatchery: www.meyerhatchery.com

Traditional hatchery specializing in three-chick minimums for backyard microflocks.

Murray McMurray Hatchery: www.mcmurrayhatchery.com

One of the biggest and oldest hatcheries in the United States, boasting a wide selection of rare breed chickens as well as domestic and exotic poultry.

Sandhill Preservation Center: www.sandhillpreservation.com

Heirloom seeds and poultry sold in the interest of preserving genetic diversity.

Index

Note: Boldface numbers indicate a photo or illustration.

A

B

C

D

E

F

G

H

Credits

Front cover: Sue McDonald/Perfect Gui

Back cover: Heath Johnson/Vaniatka

Shutterstock: Africa Studio, 151, 211; Subbotina Anna, 1 (top); Anneka, 109; Ariene Studio, 119, 180–181; Noam Armonn, 61; ART Watcharapong, 44 (top); Asillem, 45 (far right); Andi Berger, 159; Stephane Bidouze, 131; bikeriderlondon, 58, 133, 170; Barry Blackburn, 13; Ramunas Bruzas, 7; Tony Campbell, 63 (top left), 78; Felix-Emanuel Catana, 71; Soraluk Chonvanich. 86–87; Critterbiz, 48 (bottom), 178; cynoclub, 49 (top), 89; davemhuntphotography, 175; Alice Day, 52; ddsign, 210; Zanna Demcenko, 139; Paul Dempsey, 47 (far left); DenisNata, 3; deterjeg, 30; egilshay, 105; eltoro69, 68; Emilio100, 115 (top); Erni, 47 (near left); Eric Fleming Photography, 24; Stephanie Frey, 50; GreenLookPook, 173; Konstantin Gushcha, 143; Arina P Habich, 22; Caz Harris Photography, 128-129; Tracey Helmboldt, 16–17; Jiri Hera, 198–199; Horse Crazy, 45 (near left), 106, 188; Anna Hoychuk, 46 (far left), 126; debra hughes, 4, 10, 18, 38, 56, 66, 88, 112, 130, 142, 162, 182, 200; terekhov igor, 72; Imageman, 40; In Tune, 94; Marcel Jancovic, 29; janecat, 132; Wuttichai jantarak, 114, 115 (bottom); jiangdi, 202; JIL Photo, 123; Gregory Johnston, 47 (far right); Kemeo, 4 (top right); Amy Kerkemeyer, 46 (near right), 146; KieferPix, 32; The Len, 90; LoloStock, 33; Torsten Lorenz, 136; Sue McDonald, 48 (top), 190; Magdanatka, 160–161, 168; alex makarenko, 69; Mona Makela, 103 (bottom); Marbury, 91 (bottom); marcovarro, 153; mary416, 154; Mike12GPH, 44 (bottom left); debra millet, 23; Jo millington, 207; MISHELLA, 49 (bottom); Tony Moran, 5; Evgeniya Moroz, 148; Catherine Murray, 25, 54–55, 59; MurzikNata, 184; Muskoka Stock Photos, 46 (far right); mythja, 110–111, 122; Jodie Nash, 45 (near right); Aksenova Natalya, 15, 189; nimon, 157; OpopO, 74; Nataliya Ostapenko, 8–9; panbazil, 66, 70; John Panella, 18; J Parker, 45 (far left); pavla, 192; PCHT, 64–65; Catalin Petolea, 197; rck_953, 176; vladimir salman, 36–37; schankz, 100; Schubbel, 166; SGr, 67; Becky Sheridan, 172; Sherjaca, 112; Shout It Out Design, 77; smikeymikey1, 135; Smit, 81; Andrea J Smith, 44 (bottom center); Gina Smith, 62, 193; Still AB, 206; stockphoto mania, 145; Diana Taliun, 108; teekaygee, 171; Paul Tessier, 164; thieury, 11, 91 (top); Max Topchii, 56; Tsekhmister, 107; underverse, 205; Vaniatka, 1 (bottom); View Factor Images, 47 (near right), 140–141; xpixel, 98; Yellowj, 204; yevgeniy11, 130; Michael Zysman, 46 (near left)

Additional photographs: Tim Belyk/Acreage Media Solutions, 44 (bottom right); Ian L. Urquhart, 57, 63 (bottom right)

Illustrations: Kristina Mercedes Urquhart, 38, 39, 41; Shutterstock: Morphart Creation, 103 (top)

About the Author

Kristina Mercedes Urquhart writes the regular column "Fowl Language" for each issue of *Chickens* magazine as well as the "City Buzz" column in *Urban Farm* magazine. In addition to supplying articles to both the *Hobby Farms* and the *Urban Farms* websites, Kristina writes for *Taproot* magazine and has blogged for *Whole Home News*. She also supplies hand-drawn illustrations for each of her columns, and her chalkboard drawings can be found in *Crackers & Dips: More than 50 Handmade Snacks* (Chronicle Books, 2013). Kristina is the buzz behind the growing beekeeping Facebook group, *The Humble Honeybee*. After receiving her master's degree in art therapy from New York University, Kristina moved to the mountains of western North Carolina with her husband, intent on living closer to the land. With their young daughter, Kristina grows food and raises animals on their modest homestead.